Lecture Notes in Mathematics 2053

Editors:
J.-M. Morel, Cachan
B. Teissier, Paris

Steffen Fröhlich

Coulomb Frames in the Normal Bundle of Surfaces in Euclidean Spaces

Topics from Differential Geometry and Geometric Analysis of Surfaces

 Springer

Steffen Fröhlich
Johannes Gutenberg University
Department of Physics, Mathematics, Computer Sciences
Mainz, Germany

ISBN 978-3-642-29845-5   ISBN 978-3-642-29846-2 (eBook)
DOI 10.1007/978-3-642-29846-2
Springer Heidelberg New York Dordrecht London

Lecture Notes in Mathematics ISSN print edition: 0075-8434
                             ISSN electronic edition: 1617-9692

Library of Congress Control Number: 2012942080

Mathematics Subject Classification (2010): 53A07, 53A10, 35J15, 35J47, 30G20

Printed on acid-free paper

Springer is part of Springer Science+Business Media (www.springer.com)

To Emil and Mihaela

Preface

These lecture notes are intended for advanced students and young researchers with interests in the analysis of partial differential equations and differential geometry. We investigate the following problems:

- What are geometrical and analytical characteristics of two-dimensional immersions of disc-type in higher-dimensional Euclidean spaces $\mathbb{R}^n$?

- What can we state about the geometry of orthogonal unit normal frames for such surfaces, as a generalization of the classical concept of unit normal vectors?

- Are there special orthogonal unit normal frames for surfaces which are particularly useful for analytical and geometrical purposes, and how can we construct such frames?

To be more explicit, we have in mind:

- Firstly, to extend treatments on elementary differential geometry of surfaces in $\mathbb{R}^3$, as presented for example in the excellent textbooks of Bär [4], Blaschke and Leichtweiß [12], Klingenberg [80], or Kühnel [84], and to continue treatments, for example, from Brauner [14] or Eschenburg and Jost [44], where selected aspects of surface geometry in Euclidean spaces are already discussed, to an analytical theory of surfaces in Euclidean spaces together with its elements of complex analysis and partial differential equations.

- Secondly, to provide a new approach, as comprehensive as possible, to the construction of orthogonal unit normal frames for surfaces which arise from certain geometric variational problems, so-called *normal Coulomb frames,* together with its elements from the theory of non-linear elliptic systems and modern harmonic analysis.

Our lecture notes contain four chapters which are organized as follows:

- *Chapter 1: Surface geometry*

  We present a comprehensive discussion of the differential geometry of surfaces immersed in Euclidean spaces.

This, in particular, includes the definition of orthogonal unit normal frames for surfaces, one central aspect of our analysis, as well as orthogonal transitions between them.

Furthermore, we derive the differential equations of Gauß and Weingarten as well as the corresponding integrability conditions of Codazzi–Mainardi, Gauß and Ricci. Based on these fundamental identities we introduce important curvature quantities of surfaces, for example the curvature tensor of the normal bundle which plays a particular role in our considerations.

Surface geometry benefits a lot from the theory of generalized analytic functions. To give an idea of what this means we want to conclude the first chapter with proving holomorphy of the so-called Hopf vector which in particular allows us to characterize the zeros of the Gauss curvature of minimal surfaces.

- *Chapter 2: Elliptic systems*

This intermediate chapter begins by introducing the theory of non-linear elliptic systems with quadratic growth in the gradient, and then presents some results concerning curvature estimates and theorems of Bernstein-type for surfaces in Euclidean spaces of arbitrary dimensions.

A famous result of S. Bernstein states that a smooth minimal graph in $\mathbb{R}^3$, defined on the whole plane $\mathbb{R}^2$, must necessarily be a plane. Today we know various strategies to prove this result, and the idea goes back to E. Heinz to establish first a curvature estimate and to deduce Bernstein's result in a second step. However, minimal surfaces with higher codimensions do not share this Bernstein property, as one of our main examples $X(w) = (w, w^2) \in \mathbb{R}^4$ with $w = u + iv$ convincingly shows. It is still a great challenge to find geometrical criteria, preferably in terms of the curvature quantities of the surfaces' normal bundles, which guarantee the validity of Bernstein's theorem.

We must admit that we can only discuss briefly some points where we would wish to employ our tools we develop in this book, but up to now we cannot continue to drive further developments.

- *Chapter 3: Normal Coulomb frames in $\mathbb{R}^4$*

With this chapter we begin our study of constructing normal Coulomb frames for surfaces immersed in Euclidean space $\mathbb{R}^4$.

Normal Coulomb frames are critical for a new functional of total torsion. We present the associated Euler–Lagrange equation and discuss its solution via a Neumann boundary value problem. A proof of the "minimal character" of normal Coulomb frames follows immediately.

Using methods from potential theory and complex analysis we establish various analytical tools to control these special frames. For example, we present two different methods to bound their torsion (connection) coefficients. Methods from the theory of generalized analytic functions will play again an important role.

We conclude the third chapter with a class of minimal graphs for which we can explicitly compute normal Coulomb frames.

- *Chapter 4: Normal Coulomb frames in $\mathbb{R}^{n+2}$*

Now we consider two-dimensional surfaces immersed in Euclidean spaces $\mathbb{R}^{n+2}$ of arbitrary dimension. The construction of normal Coulomb frames turns out to be more intricate and requires a profound analysis of non-linear elliptic systems in two variables.

The Euler–Lagrange equations of the functional of total torsion are identified as non-linear elliptic systems with quadratic growth in the gradient, and, more exactly, the non-linearity in the gradient is of so-called *curl-type*, while the Euler–Lagrange equations appear in a *div-curl-form*.

We discuss the interplay between curvatures of the normal bundles and torsion properties of normal Coulomb frames. It turns out that such frames are free of torsion if and only if the normal bundle is flat.

Existence of normal Coulomb frames is then established by solving a variational problem in a weak sense using ideas of F. Helein [64]. This, of course, ensures minimality, but we are also interested in classical regularity of our frames. For this purpose we employ deep results of the theory of non-linear elliptic systems of div-curl-type and benefit from the work of many authors: E. Heinz, S. Hildebrandt, F. Helein, F. Müller, S. Müller, T. Rivière, F. Sauvigny, A. Schikorra, E.M. Stein, F. Tomi, H.C. Wente, and many others.

Parallel frames in the normal bundle are often studied in the literature. These are special normal Coulomb frames, namely those with vanishing torsion coefficients, and so they only exist if the normal bundle is flat. In our lecture notes we will mainly consider *non-flat normal bundles* and therefore *nonparallel normal frames*.

Parallel normal frames are widely used in physics, see for example da Costa [30] for a geometric presentation of certain physical problems in quantum mechanics or Burchard and Thomas [19] for an analytical description of the dynamics of Euler's elastic curves. The treatment of such problems in the more general context of nonparallel normal frames is surely desirable but must be left open for the future.

Many fundamental mathematical problems are also left open: How can one construct normal Coulomb frames on surfaces of higher topological type or on higher-dimensional manifolds? Is it possible to combine our results with Helein's construction of tangential Coulomb frames on surfaces from [64]? How can one construct Coulomb frames on manifolds immersed in general Riemannian spaces or Lorentzian spaces? This would surely open the door to applications in general relativity or string theory. The reader is invited to join in the discussion.

Most of the results presented here were obtained in a very fruitful collaboration with Frank Müller from the University of Duisburg-Essen. The reader finds our original approaches in [49, 50] and [51].

I would like to thank the members of Springer for their helpful collaboration, for their support and for their care in preparing this work for print.

Mainz, Germany   Steffen Fröhlich

Contents

List of Symbols

Domains of definition

B	Open unit disc
$\overline{B}$	Closed unit disc
∂B	Boundary of the unit disc
$\overline{B}_R$	Closed disc of radius R

Normal vectors

N_σ	Unit normal vector
$\mathfrak{N}$	Orthonormal normal frame (ONF)
$\mathbf{R}$	Orthogonal rotation of orthonormal normal frames

Subspaces

$\mathbb{T}_X(w)$	Tangential space at $w \in B$
$\mathbb{N}_X(w)$	Normal space at $w \in B$
$\mathcal{N}(X)$	Normal bundle

Fundamental forms and Hopf vector

g_{ij}, g^{ij}	First fundamental form
$L_{\sigma,ij}$	Second fundamental form w.r.t. N_σ
$\mathcal{H}$	Hopf vector

Connection coefficients

Γ_{ij}^k	Christoffel symbols
$T_{\sigma,i}^{\vartheta}$	Torsion coefficients
T_σ^{ϑ}	Complex-valued torsion vector
$\mathbf{T}_i$	Torsion matrix
$\mathcal{T}$	Grassmann type vector

Curvatures and curvature tensors

κ_g	Geodesic curvature
K	Gauss curvature
H	Scalar mean curvature
$\mathfrak{H}$	Mean curvature vector
R^{ℓ}_{ijk}, R_{nijk}	Riemannian curvature tensor

Normal curvatures

$S^{\vartheta}_{\sigma,ij}$	Curvature tensor of the normal bundle
S	Scalar normal curvature
S^{ω}_{σ}	Normal sectional curvature
$\mathfrak{S}$	Normal curvature vector
$\mathbf{S}_{12}$	Normal curvature matrix

Integral functions

$\tau, \tau^{(\sigma\omega)}$	Integral functions
$\mathfrak{T}$	Grassmann-type vector

Parametric functionals

$\mathscr{A}[X]$	Area functional
$\mathscr{T}[\mathfrak{N}]$	Total torsion functional

Chapter 1
Surface Geometry

Abstract We present a comprehensive discussion of the differential geometry of surfaces immersed in Euclidean spaces.

This, in particular, includes the definition of orthogonal unit normal frames for surfaces, one central aspect of our analysis, as well as orthogonal transitions between them.

Furthermore, we derive the differential equations of Gauß and Weingarten as well as the corresponding integrability conditions of Codazzi–Mainardi, Gauß and Ricci. Based on these fundamental identities we introduce important curvature quantities of surfaces, for example the curvature tensor of the normal bundle which plays a particular role in our considerations.

Surface geometry benefits a lot from the theory of generalized analytic functions. To give an idea of what this means we want to conclude the first chapter with proving holomorphy of the so-called Hopf vector which in particular allows us to characterize the zeros of the Gauss curvature of minimal surfaces.

1.1 Regular Surfaces

1.1.1 First Definitions

Let $n \geq 1$ be an integer. The main objects of our considerations are vector-valued mappings

$$X = X(u, v) = \left(x^1(u, v), \ldots, x^{n+2}(u, v)\right), \quad (u, v) \in \overline{B},$$

defined on the topological closure $\overline{B} \subset \mathbb{R}^2$ of the open unit disc

$$B := \left\{ w = (u, v) \in \mathbb{R}^2 \ : \ u^2 + v^2 < 1 \right\}.$$

S. Fröhlich, *Coulomb Frames in the Normal Bundle of Surfaces in Euclidean Spaces*,
Lecture Notes in Mathematics 2053, DOI 10.1007/978-3-642-29846-2_1,
© Springer-Verlag Berlin Heidelberg 2012

From the point of view of analysis and differential geometry we always want to assume (until not presumed otherwise)

- $X \in C^{k,\alpha}(\overline{B}, \mathbb{R}^{n+2})$ with an integer $k \geq 4$ and a Hölder exponent $\alpha \in (0, 1)$, as well as
- rank $DX(u, v) = 2$ for all $(u, v) \in \overline{B}$

for the Jacobian $DX \in \mathbb{R}^{2 \times (n+2)}$ of X, i.e. at each point $w \in \overline{B}$ there is a non-degenerate, two-dimensional tangential plane.

The mapping X thus represents a regular surface or two-dimensional immersion of disc-type.

1.1.2   Tangential Space and Normal Space

Since X represents an immersion, at each point $w \in \overline{B}$ there exist two linearly independent *tangential vectors*

$$X_u(w) \equiv \frac{\partial X(w)}{\partial u} \quad \text{and} \quad X_v(w) \equiv \frac{\partial X(w)}{\partial v} \,,$$

represented analytically by the derivatives of X,[1] which span the two-dimensional *tangential space* $\mathbb{T}_X(w)$ *at that point $w \in \overline{B}$* :

$$\mathbb{T}_X(w) := \operatorname{span}\{X_u(w), X_v(w)\} \cong \mathbb{R}^2 \,.$$

Its orthogonal complement forms the *normal space* $\mathbb{N}_X(w)$ *at $w \in \overline{B}$*, i.e.

$$\mathbb{N}_X(w) := \{Z \in \mathbb{R}^{n+2} : X_u(w) \cdot Z = X_v(w) \cdot Z = 0\} \cong \mathbb{R}^n \,,$$

where $X \cdot Y$ denotes the inner Euclidean product between two vectors $X, Y \in \mathbb{R}^{n+2}$,

$$X \cdot Y := \sum_{i=1}^{n+2} x^i y^i \,.$$

1.1.3   Orthonormal Normal Frames

At each point $w \in \overline{B}$ we may choose $n \geq 1$ unit normal vectors $N_\sigma = N_\sigma(w)$ for $\sigma = 1, \ldots, n$, satisfying the following orthogonality relations

---

[1]Symbols like X_u, $N_{\sigma,u}$, $g_{ij,v}$ etc. denote partial derivatives w.r.t. u resp. v.

$$N_\sigma \cdot N_\vartheta = \delta_{\sigma\vartheta} = \begin{cases} 1 \text{ for } \sigma = \vartheta \\ 0 \text{ for } \sigma \neq \vartheta \end{cases} \quad \text{for all } \sigma, \vartheta = 1, \ldots, n$$

with Kronecker's symbol $\delta_{\sigma\vartheta}$. We also use the notations $\delta_{\sigma\vartheta}$, δ^ϑ_σ, or $\delta^{\sigma\vartheta}$ for this symbol. Now choose the $N_\sigma(w)$ in such a way that:

(a) They span the normal space $\mathbb{N}_X(w)$ at $w \in \overline{B}$,
(b) They are oriented in the following sense

$$\det \left(X_u, X_v, N_1, \ldots, N_n \right) > 0.$$

Thanks to the contractibility of the domain $\overline{B}$ we can extend this system of orthogonal unit normal vectors in a differentiable way to the whole domain $\overline{B}$.

Definition 1.1. A system

$$\mathfrak{N} = (N_1, \ldots, N_n) \in C^{k-1,\alpha}(\overline{B}, \mathbb{R}^{n \times (n+2)}),$$

which consists of $n \geq 1$ orthogonal unit normal vectors $N_\sigma = N_\sigma(w)$, oriented in the above sense, which moves $C^{k-1,\alpha}$-smoothly along the whole surface X, and which spans the n-dimensional normal space $\mathbb{N}_X(w)$ at each $w \in \overline{B}$, is called an *orthogonal normal frame*, or shortly *ONF*.

1.2 Examples

1.2.1 Surface Graphs

We consider some important examples of surfaces.

Definition 1.2. A *surface graph* is a mapping

$$\mathbb{R}^2 \ni (x, y) \mapsto X(x, y) = \left(x, y, z_1(x, y), \ldots, z_n(x, y) \right) \in \mathbb{R}^{n+2}$$

with sufficiently smooth functions z_σ, $\sigma = 1, \ldots, n$, generating the graph.

Surface graphs are always immersions. We can specify a possible "normal frame" consisting of the unit normal vectors

$$N_1 = \frac{1}{\sqrt{1 + |\nabla z_1|^2}} \left(-z_{1,x}, -z_{1,y}, 1, 0, 0, \ldots, 0 \right),$$

$$N_2 = \frac{1}{\sqrt{1 + |\nabla z_2|^2}} \left(-z_{2,x}, -z_{2,y}, 0, 1, 0, \ldots, 0 \right) \quad \text{etc.}$$

with $z_{\sigma,x}$ and $z_{\sigma,y}$ denoting the partial derivatives of z_σ w.r.t. the coordinates x resp. y, and $\nabla z_\sigma = (z_{\sigma,x}, z_{\sigma,y}) \in \mathbb{R}^2$ is the Euclidean gradient of z_σ.

Definition 1.3. These special N_σ are called the *Euler unit normal vectors* of the graph X.

In general, Euler unit normals are not orthogonal, but by means of Gram–Schmidt orthogonalization we can always construct an orthogonal unit basis from an Euler unit normal vector frame.

1.2.2 Holomorphic Surface Graphs

Let us consider surface graphs of the special form

$$\mathbb{R}^2 \ni (x, y) \mapsto X(x, y) = \big(x, y, \varphi(x, y), \psi(x, y)\big) \in \mathbb{R}^4$$

with φ and ψ being real resp. imaginary part of a complex-valued holomorphic function

$$\Phi(x, y) = \varphi(x, y) + i\psi(x, y) \in \mathbb{C}$$

satisfying the Cauchy–Riemann equations

$$\varphi_x = \psi_y, \quad \varphi_y = -\psi_x.$$

For example,

$$\varphi + i\psi = (x + iy)^2 = x^2 - y^2 + 2ixy$$

will play a particular role in our analysis.

The associated Euler unit normal vectors of such a holomorphic graph,

$$N_1 = \frac{1}{\sqrt{1 + |\nabla\varphi|^2}} \big(-\varphi_x, -\varphi_y, 1, 0\big),$$

$$N_2 = \frac{1}{\sqrt{1 + |\nabla\psi|^2}} \big(-\psi_x, -\psi_y, 0, 1\big) = \frac{1}{\sqrt{1 + |\nabla\varphi|^2}} \big(\varphi_y, -\varphi_x, 0, 1\big),$$

form an ONF in the normal space since we immediately verify

$$N_1 \cdot N_2 = \frac{1}{1 + |\nabla\varphi|^2} \big(-\varphi_x\varphi_y + \varphi_y\varphi_x\big) = 0.$$

We will repeatedly return to holomorphic graphs in the course of our considerations, in particular when we discuss Bernstein's principle and curvature estimates for minimal surfaces in the second chapter, or in Chap. 3 when we prove that Euler unit normal frames for certain holomorphic graphs of the form $(w, \Phi(w))$ represent normal Coulomb frames—the central topic of this book.

1.2.3 The Veronese Surface

This is the surface

$$X(x, y; \lambda) = \lambda \left(\frac{yz}{\sqrt{3}}, \frac{xz}{\sqrt{3}}, \frac{xy}{\sqrt{3}}, \frac{x^2 - y^2}{2\sqrt{3}}, \frac{1}{6} \left(x^2 + y^2 - 2z^2 \right) \right)$$

with $x^2 + y^2 + z^2 = 3$, $\lambda \in \mathbb{R}$,

first described by G. Veronese. From Chen and Ludden [23] we infer the following interesting properties which will become more clear after the study of the first two chapters of our book.

Proposition 1.1. *The Veronese surface, as a compact surface (without boundary) in $\mathbb{R}^5$, has parallel mean curvature vector and constant normal curvature. Furthermore, it has constant Gauß curvature.*

A calculation shows that the parametrization $X(x, y; 1)$ maps points (x, y, z) of the two-dimensional sphere with radius $\sqrt{3}$,

$$\{(x, y, z) \in \mathbb{R}^3 : x^2 + y^2 + z^2 = 3\} \subset \mathbb{R}^3,$$

into the four-dimensional unit sphere

$$S^4 := \{(x_1, x_2, x_3, x_4, x_5) : x_1^2 + x_2^2 + x_3^2 + x_4^2 + x_5^2 = 1\} \subset \mathbb{R}^5$$

in such a way that two points (x, y, z) and $(-x, -y, -z)$ are mapped into the same point of S^4. Thus, we have a parametrization of a *real projective plane* in $\mathbb{R}^5$.

For a good introduction to the Veronese surface we refer the reader to Albrecht [1] who presents also modern applications of it in the field of Computer Aided Geometric Design.

The Veronese surface is a *minimal surface* and, as shown in Li [86], it is also a *Willmore surface* in S^4. These two properties make this surface so attractive for the geometric analysis, in particular for the famous Willmore problem.

For detailed discussions about the interplay between minimal surfaces in Euclidean spaces or in spheres and Willmore surfaces on the one hand, and methods of modern harmonic analysis on the other hand, we refer the reader i.e. to Bryant [16], Helein [64], Li [86], Rivière [99], Weiner [120], or Willmore [125]. Especially in the fourth chapter we will employ many of Helein's ideas and methods from [64] for our construction of normal Coulomb frames.

1.3   The Fundamental Forms

1.3.1   The First Fundamental Form

To apply the classical tensor calculus we introduce the notation $u^1 := u, u^2 := v$.

Definition 1.4. The coefficients g_{ij} of the *first fundamental form* $\mathbf{g} \in \mathbb{R}^{2 \times 2}$ of the immersion X are defined by

$$g_{ij} := X_{u^i} \cdot X_{u^j}, \quad i, j = 1, 2.$$

The differential line element of the surface w.r.t. this form then reads as follows

$$ds^2 = \sum_{i,j=1}^{2} g_{ij} \, du^i du^j.$$

Formally this line element results from inserting the parametric representation of $X = X(u, v)$ into the Euclidean form

$$ds^2 = \sum_{k,\ell=1}^{n+2} \delta_{k\ell} \, dx^k dx^\ell$$

of the embedding space $\mathbb{R}^{n+2}$ with Cartesian coordinates x^k for $k = 1, \ldots, n + 2$. Namely, we calculate

$$ds^2 = \sum_{k,\ell=1}^{n+2} \delta_{k\ell}(x_u^k \, du + x_v^k \, dv)(x_u^\ell \, du + x_v^\ell \, dv)$$

$$= \sum_{k=1}^{n+2} \left\{ (x_u^k)^2 \, du^2 + 2(x_u^k x_v^k) \, du dv + (x_v^k)^2 \, dv^2 \right\}.$$

Notice that the first fundamental form $\mathbf{g}$ is invertible on account of the regularity property rank $DX = 2$ for the Jacobian DX of X. For its inverse matrix we write

$$\mathbf{g}^{-1} = (g^{ij})_{i,j=1,2} \in \mathbb{R}^{2 \times 2}.$$

At each point $w \in \overline{B}$ it then holds

$$(\mathbf{g} \circ \mathbf{g}^{-1})_{ik} = \sum_{j=1}^{2} g_{ij} g^{jk} = \delta_i^k \quad \text{with Kronecker's symbol } \delta_i^k.$$

1.3.2 The Tensor of the Second Fundamental Forms

We come to

Definition 1.5. To each unit normal vector N_σ of a given ONF $\mathfrak{N} = (N_1, \ldots, N_n)$ we assign a *second fundamental form* with coefficients

$$L_{\sigma,ij} := X_{u^i u^j} \cdot N_\sigma, \quad i, j = 1, 2, \ \sigma = 1, \ldots, n.$$

Notice that

$$X_{u^i u^j} \cdot N_\sigma = -X_{u^i} \cdot N_{\sigma, u^j}$$

which follows directly after differentiation of the orthogonality relations $X_{u^i} \cdot N_\sigma = 0$ for all $i = 1, 2$ and $\sigma = 1, \ldots, n$.

In case $n = 1$ of one codimension there is only one second fundamental form which is uniquely defined up to orientation of the unit normal vector N.

1.3.3 Conformal Parameters

We will often work with *conformal parameters* $(u, v) \in \overline{B}$ satisfying the *conformality relations*

$$g_{11} = W = g_{22}, \quad g_{12} = 0 \quad \text{in } \overline{B}$$

with the area element

$$W := \sqrt{g_{11} g_{22} - g_{12}^2}.$$

This area element then represents the *conformal factor* w.r.t. the surface's conformal parametrization.

Introducing conformal parameters is justified by results like (see [107])

Proposition 1.2. *Suppose that the coefficients a, b and c of the Riemannian metric*

$$ds^2 = a \, du^2 + 2b \, du dv + c \, dv^2$$

are of class $C^{1+\alpha}(\overline{B}, \mathbb{R})$ with $\alpha \in (0, 1)$. Then there is a conformal parameter system $(u, v) \in \overline{B}$.

Recall that ds^2 is of *Riemannian type* if $ac - b^2 \geq \eta_0 > 0$ in $\overline{B}$. The regularity condition required here is satisfied in our situation due to $g_{ij} \in C^{k-1}(\overline{B})$ with $k \geq 4$.

While Sauvigny's result holds *in the large*, i.e. on the whole closed disc $\overline{B}$, another optimal result *in the small* goes back to Korn [83] and Lichtenstein [87], see also Chern [24] for a simplified proof.

Proposition 1.3. *Suppose that the coefficients a, b and c of the Riemannian metric*

$$ds^2 = a \, du^2 + 2b \, du dv + c \, dv^2$$

satisfy a Hölder condition in $\overline{B}$. *Then for every point* $w \in B$ *there exists an open neighborhood over which the surface can be parametrized conformally.*

The uniformization principle would now guarantee the *global* existence of conformal parameters, see i.e. the classical monographs Courant [28] or Nitsche [92] for comprehensive discussions.

Note also the different regularity assumptions in the propositions due to different analytical approaches. However, Hartman and Wintner in [58] pointed out that if the coefficients a, b and c are only continuous then there need not exist any parameter transformation into a conformal form. Thus, the weakest regularity conditions are given by Korn and Lichtenstein.

For recent developments on this subject, namely in connection with the classical Plateau problem for minimal surfaces, we refer the reader to Hildebrandt and von der Mosel [67] as well as to the detailed analysis in Dierkes et al. [34].

1.3.4 Example: Holomorphic Surfaces

For $w = u + iv \in \overline{B}$ we consider mappings

$$X(w) = \big(\Phi(w), \Psi(w)\big) \colon \overline{B} \longrightarrow \mathbb{C} \times \mathbb{C}$$

with complex-valued holomorphic functions $\Phi = (\varphi_1, \varphi_2)$ and $\Psi = (\psi_1, \psi_2)$. Real- and imaginary part are solutions of the Cauchy–Riemann equations

$$\varphi_{1,u} = \varphi_{2,v}, \quad \varphi_{1,v} = -\varphi_{2,u} \quad \text{and}$$

$$\psi_{1,u} = \psi_{2,v}, \quad \psi_{1,v} = -\psi_{2,u}.$$

We calculate the coefficients of the first fundamental form

$$g_{11} = X_u \cdot X_u = (\varphi_{1,u}, \varphi_{2,u}, \psi_{1,u}, \psi_{2,u})^2 = \varphi_{1,u}^2 + \varphi_{2,u}^2 + \psi_{1,u}^2 + \psi_{2,u}^2,$$

$$g_{22} = X_v \cdot X_v = \varphi_{1,v}^2 + \varphi_{2,v}^2 + \psi_{1,v}^2 + \psi_{2,v}^2 = \varphi_{2,u}^2 + \varphi_{1,u}^2 + \psi_{2,u}^2 + \psi_{1,u}^2 = g_{11},$$

$$g_{12} = X_u \cdot X_v = \varphi_{1,u}\varphi_{1,v} + \varphi_{2,u}\varphi_{2,v} + \psi_{1,u}\psi_{1,v} + \psi_{2,u}\psi_{2,v} = 0.$$

Proposition 1.4. *The map* $X(w) = (\Phi(w), \Psi(w))$, $w \in \overline{B}$, *with complex-valued, holomorphic functions* Φ *and* Ψ *is conformally parametrized.*

1.3.5 Outlook and Some Open Problems

Before we go into a detailed analysis we want to discuss briefly some questions we do not address in this book, admittedly due to our lack of knowledge, but which should definitely be approached in the future.

1. *Riemannian embedding spaces*

From the analytical and from the geometrical point of view it is of interest to consider immersions which live in general Riemannian spaces. For example, let $X: \overline{B} \rightarrow \mathcal{N}^{n+2}$ with a $(n+2)$-dimensional manifold $\mathcal{N}^{n+2}$ be equipped with a Riemannian metric $\eta_{k\ell}$ satisfying

$$\sum_{k,\ell=1}^{n+2} \eta_{k\ell}\xi^k\xi^\ell \geq \eta_0|\xi|^2 \quad \text{for all } \xi = (\xi^1, \ldots, \xi^{n+2}) \in \mathbb{R}^{n+2}$$

with some real $\eta_0 > 0$. The corresponding line element is then given by

$$ds^2 = \sum_{k,\ell=1}^{n+2} \eta_{k\ell}\, dx^k dx^\ell$$

from where we infer the induced line element of the surface,

$$ds^2 = \sum_{i,j=1}^{2}\sum_{k,\ell=1}^{n+2} \eta_{k\ell} x_{u^i}^k x_{u^j}^\ell\, du^i du^j = \sum_{i,j=1}^{2} \gamma_{ij}\, du^i du^j \quad \text{with}$$

$$\gamma_{ij} := \sum_{k,\ell=1}^{n+2} \eta_{k\ell} x_{u^i}^k x_{u^j}^\ell.$$

This form is of Riemannian type and admits a conformal parametrization.

2. *Lorentz spaces and space forms*

We would also like to work with immersions in pseudo-Euclidean spaces or general hyperbolic space forms of negative curvature, since all of them are relevant for many applications. The simplest example of such a manifold is the four-dimensional Minkowski space $\mathbb{R}^{3+1}$ with metric $(\eta_{ij})_{i,j=1,\ldots,4} = \text{diag}\,(-1, 1, 1, 1)$. A generalization of our calculations to the case of positively and negatively curved embedding manifolds will appear in future papers. We will frequently indicate such applications from physics: In Burchard and Thomas [19], for example, the authors employ parallel normal frames for curves in three-dimensional space to analyse their elastic properties. The study of elastic properties of surfaces with higher codimensions, on the other hand, would require to approach deep analytical questions concerning Willmore surfaces.

And, to mention a second example, further applications and also new problems should open up in the analysis of classical string actions in physics. In particular, the coupled *Nambu-Goto-Polyakov action* combines properties of minimizers or critical points of the area functional and the Willmore functional in higher-dimensional spaces, see e.g. Konopelchenko and Landolfi [82]. Such surfaces then form the basis of physical strings.

The reader is referred to the introductory but comprehensive monograph Zwiebach [128] who presents this theory from the point of view of classical calculus of variations and general relativity.

3. *Higher dimensional manifolds*

Various analytical problems appear when we work with higher dimensional manifolds instead of two-dimensional surfaces. This particularly concerns the theory of non-linear elliptic systems we employ in Chaps. 3 and 4. With the presentation at hand we cover only the two-dimensional situation.

4. *General vector bundles*

Orthonormal tangential frames are considered in Helein [64], and here we develop a theory of orthonormal frames in the normal bundle of surfaces. A theory for arbitrary vector bundles over manifolds is definitely desirable and will be a topic of future work.

5. *Curvature flows*

Another issue concerns geometric flows for surfaces in higher-dimensional spaces, in particular the mean curvature flow, see e.g. Ecker [41] for detailed discussions on the classical theory, or the recent paper Andrews and Baker [3] on curvature-pinched submanifolds and their evolution to spheres. The literature covers further numerous contributions on this problem, but mainly for surfaces with either mean curvature vector parallel in the normal bundle or even with flat normal bundles. It is a future aim to employ our theory of normal Coulomb frames to control geometric flows for objects with higher codimension.

It seems that there is even a lack of a satisfactorily treatment of *parallel-type surfaces* with higher codimensions. Beside the omnipresent problem of deriving manageable expressions for various curvature quantities of parallel surfaces, one would particularly be faced with the appearance of singularities in such a "process of parallel displacement," comparable with central problems from the mean curvature flow but surely a lot easier. We will briefly discuss such surfaces in Chap. 3.

1.4 Differential Equations

1.4.1 Problem Statement

The system $\{X_u, X_v, N_1, \ldots, N_n\}$ forms what we want to call a *moving $(n + 2)$-frame* for the immersion X. In the following we will quantify the rate of change of this frame under infinitesimal variations.

1.4.2 The Christoffel Symbols

To evaluate the derivatives of X we first need the following definition.

Definition 1.6. The *connection coefficients of the tangent bundle*[2] of the immersion X are the *Christoffel symbols*

---

[2]The tangent bundle is the collection $\bigcup_{w \in \overline{B}} \{w\} \times \mathbb{T}_X(w)$.

$$\Gamma_{ij}^k := \frac{1}{2} \sum_{\ell=1}^{2} g^{k\ell} \left(g_{\ell i, u^j} + g_{j\ell, u^i} - g_{ij, u^\ell} \right), \quad i, j, k = 1, 2.$$

Using conformal parameters from Sect. 1.3.3, the Christoffel symbols take the form

$$\Gamma_{11}^1 = \frac{W_u}{2W}, \quad \Gamma_{12}^1 = \Gamma_{21}^1 = \frac{W_v}{2W}, \quad \Gamma_{22}^1 = -\frac{W_u}{2W},$$

$$\Gamma_{11}^2 = -\frac{W_v}{2W}, \quad \Gamma_{12}^2 = \Gamma_{21}^2 = \frac{W_u}{2W}, \quad \Gamma_{22}^2 = \frac{W_v}{2W}$$

with the area element W. They decode the way of parallel transport of surface vector fields. Of central interest will be the connection coefficients of the normal bundle which we introduce below.

1.4.3 The Gauß Equations

With the help of the Christoffel symbols and the coefficients $L_{\sigma,ij}$ of the second fundamental form for some unit normal vector $N_\sigma \in \mathfrak{N}$ we can state

Proposition 1.5. *Let the immersion X together with an ONF $\mathfrak{N}$ be given. Then there hold the Gauß equations*

$$X_{u^i u^j} = \sum_{k=1}^{2} \Gamma_{ij}^k X_{u^k} + \sum_{\sigma=1}^{n} L_{\sigma,ij} N_\sigma \quad \text{for } i, j = 1, 2.$$

Proof. We follow Blaschke and Leichtweiß [12], Sect. 57 and evaluate the ansatz

$$X_{u^i u^j} = \sum_{k=1}^{2} a_{ij}^k X_{u^k} + \sum_{\vartheta=1}^{n} b_{\vartheta,ij} N_\vartheta$$

with functions a_{ij}^k and $b_{\vartheta,ij}$ to be determined. A first multiplication by N_ω gives

$$L_{\omega,ij} = X_{u^i u^j} \cdot N_\omega = \sum_{\vartheta=1}^{n} b_{\vartheta,ij} N_\vartheta \cdot N_\omega = \sum_{\vartheta=1}^{n} b_{\vartheta,ij} \delta_{\vartheta\omega} = b_{\omega,ij}.$$

To compute next the a_{ij}^k we multiply our ansatz by X_{u^ℓ} and arrive at

$$X_{u^i u^j} \cdot X_{u^\ell} = \sum_{k=1}^{2} a_{ij}^k g_{k\ell} =: a_{i\ell j}.$$

Note that $a_{i\ell j} = a_{j\ell i}$ due to Schwarz's lemma. We calculate

$$a_{i\ell j} = (X_{u^i} \cdot X_{u^\ell})_{u^j} - X_{u^i} \cdot X_{u^\ell u^j} = g_{i\ell,u^j} - a_{\ell ij}$$

which implies $g_{i\ell,u^j} = a_{i\ell j} + a_{\ell ij}$. Thus, we infer $g_{j\ell,u^i} + g_{\ell i,u^j} - g_{ij,u^\ell} = 2a_{i\ell j}$, and it therefore follows

$$\sum_{k=1}^{2} a_{ij}^k g_{k\ell} = \frac{1}{2}(g_{j\ell,u^i} + g_{\ell i,u^j} - g_{ij,u^\ell}).$$

Rearranging for the a_{ij}^k into the form

$$a_{ij}^m = \frac{1}{2}\sum_{\ell=1}^{2} g^{m\ell}(g_{j\ell,u^i} + g_{\ell i,u^j} - g_{ij,u^\ell})$$

proves the statement.   □

1.4.4   The Torsion Coefficients

To determine the infinitesimal variation of the unit normal vectors N_σ of some fixed ONF $\mathfrak{N}$ we need the following connection coefficients of the normal bundle.[3]

Definition 1.7. Let the immersion X together with an ONF $\mathfrak{N}$ be given. The *connection coefficients of its normal bundle* are the *torsion coefficients*

$$T_{\sigma,i}^\vartheta := N_{\sigma,u^i} \cdot N_\vartheta \quad \text{for } i = 1, 2 \text{ and } \sigma, \vartheta = 1, \ldots, n.$$

Taking $N_\sigma \cdot N_\vartheta = \delta_{\sigma\vartheta}$ into account we immediately infer

Proposition 1.6. *The torsion coefficients are skew-symmetric w.r.t. interchanging* $\sigma \leftrightarrow \vartheta$, *i.e. there always hold*

$$T_{\sigma,i}^\vartheta = -T_{\vartheta,i}^\sigma \quad \text{for all } i = 1, 2 \text{ and } \sigma, \vartheta = 1, \ldots, n.$$

In particular, $T_{\sigma,i}^\sigma \equiv 0$.

The torsion coefficients behave like tensors of rank 1 w.r.t. the lower i-index, and therefore they depend on the parametrization.

To justify the name "torsion coefficient" we consider an arc-length parametrized curve $c(s)$ in $\mathbb{R}^3$ together with the moving 3-frame $(\mathfrak{t}(s), \mathfrak{n}(s), \mathfrak{b}(s))$ consisting of

---

[3]The normal bundle is the collection $\bigcup_{w \in \overline{B}} \{w\} \times \mathbb{N}_X(w)$, see Definition 1.9 below.

the unit tangent vector $\mathfrak{t}(s)$, the unit normal vector $\mathfrak{n}(s)$ and the unit binormal vector $\mathfrak{b}(s)$. Then its curvature $\kappa(s)$ and torsion $\tau(s)$ are given by

$$\kappa(s) = |\mathfrak{t}'(s)|, \quad \tau(s) = \mathfrak{n}'(s) \cdot \mathfrak{b}(s),$$

and this already clarifies the analogy to our definition of the torsion coefficients.

In fact, it was Weyl in [124] who first used the terminology "torsion": *Aus einem normalen Vektor* **n** *in* P *entsteht ein Vektor* **n**$'$ + $d\mathbf{t}$ (**n**$'$ *normal*, $d\mathbf{t}$ *tangential*). *Die infinitesimale lineare Abbildung* **n** $\rightarrow$ **n**$'$ *von* $\mathfrak{N}_P$ *auf* $\mathfrak{N}_{P'}$ *ist die Torsion*.[4]

1.4.5 The Weingarten Equations

We determine the variation of the unit normal vectors N_σ of a given ONF $\mathfrak{N}$.

Proposition 1.7. *Let the immersion* X *together with an ONF* $\mathfrak{N}$ *be given. Then there hold the Weingarten equations*

$$N_{\sigma,u^i} = -\sum_{j,k=1}^{2} L_{\sigma,ij} \, g^{jk} X_{u^k} + \sum_{\vartheta=1}^{n} T_{\sigma,i}^{\vartheta} N_\vartheta$$

for $i = 1, 2$ *and* $\sigma = 1, \ldots, n$.

Proof. We follow Blaschke and Leichtweiß [12] again and determine the unknown functions $a_{\sigma,i}^k$ and $b_{\sigma,i}^\vartheta$ of the ansatz

$$N_{\sigma,u^i} = \sum_{k=1}^{2} a_{\sigma,i}^k X_{u^k} + \sum_{\vartheta=1}^{n} b_{\sigma,i}^\vartheta N_\vartheta \, .$$

Multiplication by X_{u^ℓ} gives

$$-L_{\sigma,i\ell} = N_{\sigma,u^i} \cdot X_{u^\ell} = \sum_{k=1}^{2} a_{\sigma,i}^k X_{u^k} \cdot X_{u^\ell} = \sum_{k=1}^{2} a_{\sigma,i}^k g_{k\ell}$$

and therefore

$$a_{\sigma,i}^m = -\sum_{\ell=1}^{2} L_{\sigma,i\ell} \, g^{\ell m} \, .$$

---

[4]From a normal vector **n** in P there arises a vector **n**$'$ + $d\mathbf{t}$ (**n**$'$ normal, $d\mathbf{t}$ tangential). The infinitesimal linear mapping **n** $\rightarrow$ **n**$'$ from $\mathfrak{N}_P$ into $\mathfrak{N}_{P'}$ is the torsion; see the next paragraph.

A second multiplication by N_ω shows

$$T_{\sigma,i}^\omega = N_{\sigma,u^i} \cdot N_\omega = \sum_{\vartheta=1}^n b_{\sigma,i}^\vartheta N_\vartheta \cdot N_\omega = \sum_{\vartheta=1}^n b_{\sigma,i}^\vartheta \delta_{\vartheta\omega} = b_{\sigma,i}^\omega$$

proving the statement. $\square$

1.5 Integrability Conditions

1.5.1 Problem Statement

In view of $X \in C^4(\overline{B}, \mathbb{R}^{n+2})$ there hold various integrability conditions which we present in this section.

In particular, a differentiation of the Gauß equations gives us necessary conditions for the third derivatives of the surface vector X resp. the second derivatives of the tangential vectors X_{u^i} in form of the Codazzi–Mainardi equations and the famous theorema egregium:

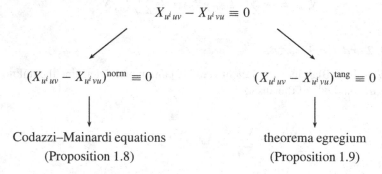

$$X_{u^i uv} - X_{u^i vu} \equiv 0$$

$$(X_{u^i uv} - X_{u^i vu})^{\text{norm}} \equiv 0 \qquad (X_{u^i uv} - X_{u^i vu})^{\text{tang}} \equiv 0$$

Codazzi–Mainardi equations   theorema egregium
(Proposition 1.8)   (Proposition 1.9)

where the upper "norm" and "tang" mean the normal resp. tangential components.

The Codazzi–Mainardi equations also arise from a differentiation of the Weingarten equations, and this latter system is finally the source of a third system of integrability conditions, named after Ricci, which has no counter-part in case $n = 1$ of one codimension.

Concerning the historical background we would like to remark that the theorema egregium and the Codazzi–Mainardi equations for general submanifolds were first derived by Voss in [118], while the Ricci equations are actually named after Ricci 1888. The reader finds more references e.g. in the survey articles Chen [22], Kobayashi [81], or in the classical monographs Eisenhart [43], Schouten [110].

Analogously we will proceed with differentiating the Weingarten equations to get

$$N_{\sigma,uv} - N_{\sigma,vu} \equiv 0$$

$$(N_{\sigma,uv} - N_{\sigma,vu})^{\text{tang}} \equiv 0 \qquad\qquad (N_{\sigma,uv} - N_{\sigma,vu})^{\text{norm}} \equiv 0$$

Codazzi–Mainardi equations   Ricci equations
(Proposition 1.8)   (Proposition 1.11)

To be precise, let us now start with the Gauß equations from Sect. 1.4.3, i.e.

$$X_{u^i u} = \Gamma_{i1}^1 X_u + \Gamma_{i1}^2 X_v + \sum_{\sigma=1}^n L_{\sigma,i1} N_\sigma \,,$$

$$X_{u^i v} = \Gamma_{i2}^1 X_u + \Gamma_{i2}^2 X_v + \sum_{\sigma=1}^n L_{\sigma,i2} N_\sigma$$

for $i = 1, 2$. We differentiate the first equation w.r.t. v,[5]

$$X_{u^i uv} = \left\{ \partial_v \Gamma_{i1}^1 + \Gamma_{i1}^1 \Gamma_{12}^1 + \Gamma_{i1}^2 \Gamma_{22}^1 - \sum_{\ell=1}^2 \sum_{\sigma=1}^n L_{\sigma,i1} L_{\sigma,2\ell} g^{\ell 1} \right\} X_u$$

$$+ \left\{ \partial_v \Gamma_{i1}^2 + \Gamma_{i1}^1 \Gamma_{12}^2 + \Gamma_{i1}^2 \Gamma_{22}^2 - \sum_{\ell=1}^2 \sum_{\sigma=1}^n L_{\sigma,i1} L_{\sigma,2\ell} g^{\ell 2} \right\} X_v$$

$$+ \sum_{\omega=1}^n \left\{ \partial_v L_{\omega,i1} + \Gamma_{i1}^1 L_{\omega,12} + \Gamma_{i1}^2 L_{\omega,22} + \sum_{\sigma=1}^n L_{\sigma,i1} T_{\sigma,2}^\omega \right\} N_\omega \,,$$

and the second equation w.r.t. u,

$$X_{u^i vu} = \left\{ \partial_u \Gamma_{i2}^1 + \Gamma_{i2}^1 \Gamma_{11}^1 + \Gamma_{i2}^2 \Gamma_{12}^1 - \sum_{\ell=1}^2 \sum_{\sigma=1}^n L_{\sigma,i2} L_{\sigma,1\ell} g^{\ell 1} \right\} X_u$$

$$+ \left\{ \partial_u \Gamma_{i2}^2 + \Gamma_{i2}^1 \Gamma_{11}^2 + \Gamma_{i2}^2 \Gamma_{12}^2 - \sum_{\ell=1}^2 \sum_{\sigma=1}^n L_{\sigma,i2} L_{\sigma,1\ell} g^{\ell 2} \right\} X_v$$

$$+ \sum_{\omega=1}^n \left\{ \partial_u L_{\omega,i2} + \Gamma_{i2}^1 L_{\omega,11} + \Gamma_{i2}^2 L_{\omega,12} + \sum_{\sigma=1}^n L_{\sigma,i2} T_{\sigma,1}^\omega \right\} N_\omega \,.$$

---

[5]For a better overview and to avoid too many commas we also use the symbol $\partial_{u^i} \Gamma_{i1}^1 := \Gamma_{i1,u^i}^1$ etc. for the partial derivatives.

Comparing the tangential and normal parts of these two identities gives the first set of our integrability conditions as follows.

1.5.2   The Integrability Conditions of Codazzi and Mainardi

Namely, from the identity $(X_{u^i uv} - X_{u^i vu})^{\text{norm}} \equiv 0$ we particularly infer the following result (we interchange σ and ω).

Proposition 1.8. *Let the immersion* X *together with an ONF* $\mathfrak{N}$ *be given. Then*

$$\partial_v L_{\sigma,i1} + \Gamma_{i1}^1 L_{\sigma,12} + \Gamma_{i1}^2 L_{\sigma,22} + \sum_{\omega=1}^{n} L_{\omega,i1} T_{\omega,2}^{\sigma}$$

$$= \partial_u L_{\sigma,i2} + \Gamma_{i2}^1 L_{\sigma,11} + \Gamma_{i2}^2 L_{\sigma,12} + \sum_{\omega=1}^{n} L_{\omega,i2} T_{\omega,1}^{\sigma}$$

for $i = 1, 2$ *and* $\sigma = 1, 2, \ldots, n$.

In contrast to the case $n = 1$ of one codimension, i.e. for surfaces in $\mathbb{R}^3$, these equations contain the torsion coefficients introduced in Sect. 1.4.4, and this indicates the richer geometry in the higher-codimensional situation.

1.5.3   The Integrability Conditions of Gauß

Next, from $(X_{u^i uv} - X_{u^i vu})^{\text{tang}} \equiv 0$ we infer

Proposition 1.9. *Let the immersion* X *together with an ONF* $\mathfrak{N}$ *be given. Then there hold the integrability conditions of Gauß*

$$\partial_v \Gamma_{i1}^\ell - \partial_u \Gamma_{i2}^\ell + \sum_{m=1}^{2} \Gamma_{i1}^m \Gamma_{m2}^\ell - \sum_{m=1}^{2} \Gamma_{i2}^m \Gamma_{m1}^\ell = \sum_{m=1}^{2} \sum_{\sigma=1}^{n} (L_{\sigma,i1} L_{\sigma,2m} - L_{\sigma,i2} L_{\sigma,1m}) g^{m\ell}$$

for $i, \ell = 1, 2$.

These equations actually do not contain the torsion coefficients. Rather they belong to the *inner geometry* of the surface as will become more clear in the next paragraphs. In case $n = 1$ they read

$$\partial_v \Gamma_{i1}^\ell - \partial_u \Gamma_{i2}^\ell + \sum_{m=1}^{2}(\Gamma_{i1}^m \Gamma_{m2}^\ell - \Gamma_{i2}^m \Gamma_{m1}^\ell) = \sum_{m=1}^{2}(L_{i1}L_{2m} - L_{i2}L_{1m})g^{m\ell}$$

with $L_{ij} = X_{u^i u^j} \cdot N$ denoting the coefficients of the second fundamental form w.r.t. the one unit normal vector N.

1.5.4 The Curvature Tensor of the Tangent Bundle

The left hand side of the Gauß integrability conditions gives reason to our next

Definition 1.8. The *curvature tensor of the tangent bundle* of the immersion X, also called the *Riemannian curvature tensor*, is given by components

$$R_{ijk}^\ell := \partial_{u^k} \Gamma_{ij}^\ell - \partial_{u^j} \Gamma_{ik}^\ell + \sum_{m=1}^{2}(\Gamma_{ij}^m \Gamma_{mk}^\ell - \Gamma_{ik}^m \Gamma_{mj}^\ell)$$

for $i, j, k, \ell = 1, 2$.
Its *covariant components* are

$$R_{nijk} = \sum_{\ell=1}^{2} R_{ijk}^\ell g_{\ell n} \,.$$

In our case of two dimensions for X, these R_{nijk} reduce to one essential component:

$$R_{1111} = 0, \quad R_{2222} = 0, \quad R_{1222} = 0, \quad R_{2111} = 0, \quad R_{2221} = 0, \quad R_{1112} = 0,$$

$$R_{1122} = 0, \quad R_{2211} = 0, \quad R_{1121} = 0, \quad R_{1211} = 0, \quad R_{2212} = 0, \quad R_{2122} = 0,$$

$$R_{2112} = R_{1221} = -R_{2121} = -R_{1212} \,.$$

Let us consider this fact from another direction.

1.5.5 Gauß Curvature and the Theorema Egregium

Namely, the essential component R_{2112} of the Riemannian curvature tensor represents the inner curvature of the surface X.

Proposition 1.10. *Let the immersion X together with an ONF $\mathfrak{N}$ be given. Then it holds the theorema egregium*

$$R_{2112} = KW^2$$

with the Gaussian curvature of the surface, defined by

$$K := \sum_{\sigma=1}^{n} K_{\sigma} \quad \text{with } K_{\sigma} := \frac{L_{\sigma,11}L_{\sigma,22} - L_{\sigma,12}^2}{W^2}$$

and the area element W.

Proof. Using the Gauß integrability conditions we compute

$$R_{2112} = \sum_{\ell=1}^{2} R_{112}^{\ell} g_{\ell 2} = \sum_{\ell,m=1}^{2}\sum_{\sigma=1}^{n}(L_{\sigma,11}L_{\sigma,2m} - L_{\sigma,12}L_{\sigma,1m})g^{m\ell}g_{\ell 2}$$

$$= \sum_{\ell=1}^{2}\sum_{\sigma=1}^{n}(L_{\sigma,11}L_{\sigma,22} - L_{\sigma,12}L_{\sigma,12})g^{2\ell}g_{\ell 2} = \sum_{\sigma=1}^{n} K_{\sigma}W^2 = KW^2,$$

and the statement follows.   □

The Gauss curvature K does neither depend on the choice of the parametrization $(u, v) \in \overline{B}$ nor on the choice of the ONF $\mathfrak{N}$ (of course, its components K_{σ} are not invariant; we skip a proof of these invariance properties, but for similar calculations which show the effect of rotational mappings see Sect. 1.6.4 ff.). It can be computed from the knowledge of the first fundamental form and its first and second derivatives by means of the Riemannian curvature tensor. This is the content of the famous *theorema egregium*.

1.5.6   The Integrability Conditions of Ricci

Now we want to derive Ricci's integrability conditions which have, as mentioned above, no counterpart in case $n = 1$ of one codimension.

Let us first compute the second derivatives of the unit normal vectors of some given ONF $\mathfrak{N}$:

$$N_{\sigma,uv} = -\sum_{j,k=1}^{2} \partial_v L_{\sigma,1j}\, g^{jk} X_{u^k} - \sum_{j,k=1}^{2} L_{\sigma,1j}\, \partial_v g^{jk} X_{u^k} - \sum_{j,k=1}^{2} L_{\sigma,1j}\, g^{jk} X_{u^k v}$$

$$+ \sum_{\omega=1}^{n} \partial_v T_{\sigma,1}^{\omega} N_{\omega} + \sum_{\omega=1}^{n} T_{\sigma,1}^{\omega} N_{\omega,v}$$

$$= -\sum_{j,k=1}^{2}\left\{ \partial_v L_{\sigma,1j}\, g^{jk} + L_{\sigma,1j}\, \partial_v g^{jk} + \sum_{m=1}^{2} L_{\sigma,1j}\, g^{jm} \Gamma_{m2}^{k} \right\} X_{u^k}$$

$$- \sum_{\vartheta=1}^{n} T_{\sigma,1}^{\vartheta} L_{\vartheta,2j} g^{jk} X_{u^k}$$

$$- \sum_{\omega=1}^{n} \left\{ \sum_{j,k=1}^{2} L_{\sigma,1j} g^{jk} L_{\omega,k2} - \partial_v T_{\sigma,1}^{\omega} - \sum_{\vartheta=1}^{n} T_{\sigma,1}^{\vartheta} T_{\vartheta,2}^{\omega} \right\} N_{\omega}$$

as well as

$$N_{\sigma,vu} = - \sum_{j,k=1}^{2} \left\{ \partial_u L_{\sigma,2j} g^{jk} + L_{\sigma,2j} \partial_u g^{jk} + \sum_{m=1}^{2} L_{\sigma,2j} g^{jm} \Gamma_{m1}^{k} \right\} X_{u^k}$$

$$- \sum_{\vartheta=1}^{n} T_{\sigma,2}^{\vartheta} L_{\vartheta,1j} g^{jk} X_{u^k}$$

$$- \sum_{\omega=1}^{n} \left\{ \sum_{j,k=1}^{2} L_{\sigma,2j} g^{jk} L_{\omega,k1} - \partial_u T_{\sigma,2}^{\omega} - \sum_{\vartheta=1}^{n} T_{\sigma,2}^{\vartheta} T_{\vartheta,1}^{\omega} \right\} N_{\omega}$$

using the Gauß equations and the Weingarten equations. It holds necessarily

$$N_{\sigma,uv} - N_{\sigma,vu} \equiv 0 \quad \text{for all } \sigma = 1, \ldots, n.$$

A comparison between the tangential parts would again yield the Codazzi–Mainardi equations, and thus we drop these calculations.

Notice furthermore that the condition $(N_{\sigma,uv} - N_{\sigma,vu})^{\text{norm}} \equiv 0$ is trivially satisfied in case $n = 1$ of one codimension due to the symmetry of the g^{jk} and the coefficients L_{ij} of the second fundamental form:

$$T_{\sigma,i}^{\vartheta} \equiv 0 \quad \text{and} \quad \sum_{j,k=1}^{2} L_{1j} g^{jk} L_{k2} = \sum_{j,k=1}^{2} L_{2j} g^{jk} L_{k1}.$$

But in the general case of higher codimension these conditions are not trivial, rather they yield new equations called the *Ricci integrability conditions*.

Proposition 1.11. *Let the immersion X together with an ONF $\mathfrak{N}$ be given. Then there hold the Ricci equations*

$$\partial_u T_{\sigma,2}^{\omega} - \partial_v T_{\sigma,1}^{\omega} + \sum_{\vartheta=1}^{n} (T_{\sigma,2}^{\vartheta} T_{\vartheta,1}^{\omega} - T_{\sigma,1}^{\vartheta} T_{\vartheta,2}^{\omega})$$

$$= \sum_{j,k=1}^{2} (L_{\sigma,2j} L_{\omega,k1} - L_{\sigma,1j} L_{\omega,k2}) g^{jk}$$

for $\sigma, \omega = 1, \ldots, n$.

Both sides of these identities clearly vanish identically if $n = 1$.

We will employ this Ricci integrability conditions on several occasions: For example, when we prove invariance of the normal sectional curvatures S_σ^ω in Proposition 1.12, in Proposition 2.4 when we give an upper bound of the S_σ^ω in terms of the mean and Gaussian curvature, or in Proposition 3.3 when we consider evolute type surfaces and the curvatures of their normal bundles.

1.6   The Curvature of the Normal Bundle

1.6.1   Problem Statement

In the same way as we derived the Riemannian curvature tensor from the Gauß integrability conditions we now proceed with deriving the curvature tensor of the normal bundle from the Ricci integrability conditions. In this section we give a detailed introduction to this curvature quantity.

Definition 1.9. The *normal bundle* of the immersion X is given by

$$\mathscr{N}(X) = \bigcup_{w \in \overline{B}} \{w\} \times \mathbb{N}_X(w).$$

Here are some simple examples:

1. The normal bundle of a surface in $\mathbb{R}^3$ is the collection of all normal lines, thus it resembles the so-called Grassmann manifold $G_{3,1}$.
2. Tubular neighborhoods of curves or surfaces are resembled by its normal bundle, see also the parallel type surfaces in the next chapter.

In case of higher codimension the normal bundle possesses its own non-trivial geometry so that we can assign a *curvature* to the normal bundle. If this curvature vanishes identically then the normal bundle is called *flat*.

But, for example, the holomorphic graph $X(w) = (w, w^2)$ with $w = u + iv$ has non-flat normal bundle. Developing handy analytical methods to describe curved normal bundles is our main concern.

1.6.2   The Curvature Tensor of the Normal Bundle

The following definition concerns a central notion of our considerations.

Definition 1.10. The *curvature tensor of the normal bundle* of the immersion X is given by components

$$S_{\sigma,ij}^{\omega} := \partial_{u^j} T_{\sigma,i}^{\omega} - \partial_{u^i} T_{\sigma,j}^{\omega} + \sum_{\vartheta=1}^{n} \left(T_{\sigma,i}^{\vartheta} T_{\vartheta,j}^{\omega} - T_{\sigma,j}^{\vartheta} T_{\vartheta,i}^{\omega} \right)$$

$$= \sum_{m,n=1}^{2} (L_{\sigma,im} L_{\omega,jn} - L_{\sigma,jm} L_{\omega,in}) g^{mn}.$$

Note that the second identity follows from the integrability conditions of Ricci. The $S_{\sigma,ij}^{\omega}$ now take the role of the R_{ijk}^{ℓ}. Without proof we remark that the $S_{\sigma,ij}^{\omega}$ behave like a tensor of rank 2 (w.r.t. i and j) under regular parameter transformations, and so they are neither invariant w.r.t. parameter transformations nor on rotations of the ONF, see below. Furthermore it is sufficient to focus on the components $S_{\sigma,12}^{\omega}$ because all other components vanish or are equal to these $S_{\sigma,12}^{\omega}$ up to sign.

Using conformal parameters we arrive at the representations

$$S_{\sigma,12}^{\omega} = \partial_v T_{\sigma,1}^{\omega} - \partial_u T_{\sigma,2}^{\omega} + \sum_{\vartheta=1}^{n} (T_{\sigma,1}^{\vartheta} T_{\vartheta,2}^{\omega} - T_{\sigma,2}^{\vartheta} T_{\vartheta,1}^{\omega})$$

$$= \frac{1}{W} (L_{\sigma,11} - L_{\sigma,22}) L_{\omega,12} - \frac{1}{W} (L_{\omega,11} - L_{\omega,22}) L_{\sigma,12}.$$

A discussion on curvatures for general connections can be found e.g. in Helein [64].

1.6.3 The Case $n = 2$

The definition of $S_{\sigma,12}^{\vartheta}$ takes a special form in the case $n = 2$:

$$S_{1,12}^{2} = \partial_v T_{1,1}^{2} - \partial_u T_{1,2}^{2} + T_{1,1}^{1} T_{1,2}^{2} + T_{1,1}^{2} T_{2,2}^{2} - T_{1,2}^{1} T_{1,1}^{2} - T_{1,2}^{2} T_{2,1}^{2}$$

$$= \operatorname{div} (-T_{1,2}^{2}, T_{1,1}^{2})$$

where only $\sigma = 1$ and $\vartheta = 2$ must be taken into account. Here we use the Euclidean divergence operator

$$\operatorname{div} (-T_{1,2}^{2}, T_{1,1}^{2}) = -\partial_u T_{1,2}^{2} + \partial_v T_{1,1}^{2}.$$

Furthermore, $S_{1,12}^{2}$ *does not depend on the choice of the ONF* $\mathfrak{N}$ what turns out as a by-product in the next paragraph, and finally, $W^{-1} S_{1,12}^{2}$ is even *parameter invariant*.

Thus, we are lead to the following

Definition 1.11. The *normal curvature* of the immersion $X: \overline{B} \to \mathbb{R}^4$ is given by

$$S := \frac{1}{W} S_{1,12}^{2} = \frac{1}{W} \operatorname{div} (-T_{1,2}^{2}, T_{1,1}^{2}).$$

This quantity now belongs to the inner geometry of the surface. The general situation of higher codimension is considered next.

1.6.4   The Normal Sectional Curvature

If $n > 2$ then the components $S^\omega_{\sigma,12}$ actually depend on the choice of the normal frame. So let us fix an index pair $(\sigma, \omega) \in \{1, \ldots, n\} \times \{1, \ldots, n\}$.

Proposition 1.12. *The quantity $S^\omega_{\sigma,12}$ is invariant w.r.t. rotations of the unit normal frame $\{N_\sigma, N_\omega\}$ spanning the plane $\mathscr{E} = \mathrm{span}\, \{N_\sigma, N_\omega\}$, i.e. under $SO(2)$-regular mappings of the form*

$$\widetilde{N}_\sigma = \cos\varphi N_\sigma + \sin\varphi N_\omega\,, \quad \widetilde{N}_\omega = -\sin\varphi N_\sigma + \cos\varphi N_\omega\,.$$

Proof. For the proof we use conformal parameters $(u, v) \in \overline{B}$.[6] First, with the new coefficients $\widetilde{L}_{\sigma,ij} = \widetilde{N}_\sigma \cdot X_{u^i u^j} = -\widetilde{N}_{\sigma,u^i} \cdot X_{u^j}$, and taking the Ricci integrability conditions into account, we compute

$$W\widetilde{S}^\omega_{\sigma,12} = (\widetilde{L}_{\sigma,11}\widetilde{L}_{\omega,12} - \widetilde{L}_{\sigma,21}\widetilde{L}_{\omega,11}) + (\widetilde{L}_{\sigma,12}\widetilde{L}_{\omega,22} - \widetilde{L}_{\sigma,22}\widetilde{L}_{\omega,21})$$

$$= (\cos\varphi\, L_{\sigma,11} + \sin\varphi\, L_{\omega,11})(-\sin\varphi\, L_{\sigma,12} + \cos\varphi\, L_{\omega,12})$$

$$- (\cos\varphi\, L_{\sigma,21} + \sin\varphi\, L_{\omega,21})(-\sin\varphi\, L_{\sigma,11} + \cos\varphi\, L_{\omega,11})$$

$$+ (\cos\varphi\, L_{\sigma,12} + \sin\varphi\, L_{\omega,12})(-\sin\varphi\, L_{\sigma,22} + \cos\varphi\, L_{\omega,22})$$

$$- (\cos\varphi\, L_{\sigma,22} + \sin\varphi\, L_{\omega,22})(-\sin\varphi\, L_{\sigma,21} + \cos\varphi\, L_{\omega,21})\,.$$

Collecting and evaluating all the trigonometric squares gives

$$W\widetilde{S}^\omega_{\sigma,12} = (L_{\sigma,11} - L_{\sigma,22})L_{\omega,12} - (L_{\omega,11} - L_{\omega,22})L_{\sigma,12} = WS^\omega_{\sigma,12}$$

proving the statement.   □

We make the following[7]

Definition 1.12. The invariant quantity

$$S^\omega_\sigma := \frac{1}{W}\, S^\omega_{\sigma,12}$$

is called the *normal sectional curvature* of X w.r.t. the plane $\mathscr{E} = \mathrm{span}\, \{N_\sigma, N_\omega\}$.

In the special case $n = 2$ there is only one normal sectional curvature S, and this quantity is independent of the choice of the ONF $\mathfrak{N}$.

---

[6]Note that the $S^\omega_{\sigma,12}$ differ eventually by a Jacobian after a parameter transformation.

[7]For the transformation from a given ONF $\mathfrak{N}$ to another ONF $\widetilde{\mathfrak{N}}$ we only admit mappings of class $SO(n)$ as described above.

1.6.5   Preparing the Normal Curvature Vector: Curvature Matrices

Next we set

$$\mathbf{T}_i := (T^{\vartheta}_{\sigma,i})_{\sigma,\vartheta=1,\ldots,n} \in \mathbb{R}^{n\times n}, \quad \mathbf{S}_{12} := (S^{\vartheta}_{\sigma,12})_{\sigma,\vartheta=1,\ldots,n} \in \mathbb{R}^{n\times n}.$$

We consider rotations

$$\mathbf{R} = (R^{\vartheta}_{\sigma})_{\sigma,\vartheta=1,\ldots,n} \in C^{k-1,\alpha}(\overline{B}, SO(n))$$

as special orthogonal mappings in the normal space which transform a given ONF $\mathfrak{N}$ into a new $\widetilde{\mathfrak{N}}$ by means of

$$\widetilde{N}_{\sigma} = \sum_{\vartheta=1}^{n} R^{\vartheta}_{\sigma} N_{\vartheta} \quad \text{for } \sigma = 1,\ldots,n.$$

Lemma 1.1. *There holds the transformation rule*

$$\widetilde{\mathbf{S}}_{12} = \mathbf{R} \circ \mathbf{S}_{12} \circ \mathbf{R}^t$$

with $\mathbf{R}^t$ denoting the transposition of $\mathbf{R}$.

Proof. For the proof we consider

$$\widetilde{T}^{\vartheta}_{\sigma,i} = \widetilde{N}_{\sigma,u^i} \cdot \widetilde{N}_{\vartheta} = \sum_{\alpha=1}^{n} \left(R^{\alpha}_{\sigma,u^i} N_{\alpha} + R^{\alpha}_{\sigma} N_{\alpha,u^i} \right) \cdot \sum_{\beta=1}^{n} R^{\beta}_{\vartheta} N_{\beta}$$

$$= \sum_{\alpha,\beta=1}^{n} \left(R^{\alpha}_{\sigma,u^i} R^{\beta}_{\vartheta} \delta_{\alpha\beta} + R^{\alpha}_{\sigma} R^{\beta}_{\vartheta} T^{\beta}_{\alpha,i} \right) = \sum_{\alpha=1}^{n} R^{\alpha}_{\sigma,u^i} (R^{\vartheta}_{\alpha})^t + \sum_{\alpha,\beta=1}^{n} R^{\alpha}_{\sigma} T^{\beta}_{\alpha,i} (R^{\vartheta}_{\beta})^t$$

with the agreement $(R^{\alpha}_{\vartheta})_{\vartheta,\alpha=1,\ldots,n} = (R^{\vartheta}_{\alpha})^t_{\alpha,\vartheta=1,\ldots,n}$. Thus, we arrive at the rule

$$\widetilde{\mathbf{T}}_i = \mathbf{R}_{u^i} \circ \mathbf{R}^t + \mathbf{R} \circ \mathbf{T}_i \circ \mathbf{R}^t.$$

Using this formula we evaluate (notice $\mathbf{T}_i = -\mathbf{T}^t_i$)

$$\widetilde{\mathbf{S}}_{12} = \widetilde{\mathbf{T}}_{1,v} - \widetilde{\mathbf{T}}_{2,u} - \widetilde{\mathbf{T}}_1 \circ \widetilde{\mathbf{T}}^t_2 + \widetilde{\mathbf{T}}_2 \circ \widetilde{\mathbf{T}}^t_1.$$

Namely, first

$$\widetilde{T}_{1,v} - \widetilde{T}_{2,u} = (R_u \circ R^t + R \circ T_1 \circ R^t)_v - (R_v \circ R^t + R \circ T_2 \circ R^t)_u$$
$$= R_u \circ R_v^t - R_v \circ R_u^t + R \circ (T_{1,v} - T_{2,u}) \circ R^t$$
$$+ R_v \circ T_1 \circ R^t + R \circ T_1 \circ R_v^t - R_u \circ T_2 \circ R^t - R \circ T_2 \circ R_u^t,$$

and furthermore

$$\widetilde{T}_1 \circ \widetilde{T}_2^t - \widetilde{T}_2 \circ \widetilde{T}_1^t = (R_u \circ R^t + R \circ T_1 \circ R^t) \circ (R \circ R_v^t + R \circ T_2^t \circ R^t)$$
$$- (R_v \circ R^t + R \circ T_2 \circ R^t) \circ (R \circ R_u^t + R \circ T_1^t \circ R^t)$$
$$= R_u \circ R_v^t + R_u \circ T_2^t \circ R^t + R \circ T_1 \circ R_v^t + R \circ T_1 \circ T_2^t \circ R^t$$
$$- R_v \circ R_u^t - R_v \circ T_1^t \circ R^t - R \circ T_2 \circ R_u^t - R \circ T_2 \circ T_1^t \circ R^t$$

since $R \circ R^t = R^t \circ R = \mathbb{E}^n$ with the n-dimensional unit matrix $\mathbb{E}^n$. Taking both identities together gives us

$$\widetilde{T}_{1,v} - \widetilde{T}_{2,u} - \widetilde{T}_1 \circ \widetilde{T}_2^t + \widetilde{T}_2 \circ \widetilde{T}_1^t$$
$$= R \circ (T_{1,v} - T_{2,u} - T_1 \circ T_2^t + T_2 \circ T_1^t) \circ R^t$$
$$+ R_v \circ T_1 \circ R^t + R \circ T_1 \circ R_v^t - R_u \circ T_2 \circ R^t - R \circ T_2 \circ R_u^t$$
$$- R_u \circ T_2^t \circ R^t - R \circ T_1 \circ R_v^t + R_v \circ T_1^t \circ R^t + R \circ T_2 \circ R_u^t$$
$$= R \circ (T_{1,v} - T_{2,u} - T_1 \circ T_2^t + T_2 \circ T_1^t) \circ R^t$$

using again $T_i = -T_i^t$. This proves the statement.   □

1.6.6 Preparing the Normal Curvature Vector: The Exterior Product

For the following algebraic concepts of Grassmann geometry we refer to Cartan [20] or Heil [60].

Definition 1.13. The *exterior product*

$$\wedge : \mathbb{R}^n \times \mathbb{R}^n \longrightarrow \mathbb{R}^N, \quad N = \binom{n}{2} = \frac{n(n-1)}{2},$$

can be introduced by means of the following rules:

(E1) The mapping $\mathbb{R}^n \times \mathbb{R}^n \ni (X, Y) \mapsto X \wedge Y \in \mathbb{R}^N$ is bilinear, i.e. it holds

$$(\alpha_1 X_1 + \alpha_2 X_2) \wedge (\beta_1 Y_1 + \beta_2 Y_2)$$
$$= \alpha_1 \beta_2 X_1 \wedge Y_1 + \alpha_1 \beta_2 X_1 \wedge Y_2 + \alpha_2 \beta_1 X_2 \wedge Y_1 + \alpha_2 \beta_2 X_2 \wedge Y_2$$

for all $\alpha_1, \alpha_2, \beta_1, \beta_2 \in \mathbb{R}$ and $X_1, X_2, Y_1, Y_2 \in \mathbb{R}^n$.
And $\wedge$ is skew-symmetric,

$$X \wedge Y = -Y \wedge X$$

for all $X, Y \in \mathbb{R}^n$; in particular, it holds $X \wedge X = 0$.
(E2) Let $e_1 = (1, 0, 0, \ldots, 0) \in \mathbb{R}^n$, $e_2 = (0, 1, 0, \ldots, 0) \in \mathbb{R}^n$ etc. be the standard orthonormal basis in $\mathbb{R}^n$. Then we define

$$e_1 \wedge e_2 := (1, 0, 0, \ldots, 0, 0) \in \mathbb{R}^N,$$
$$e_1 \wedge e_3 := (0, 1, 0, \ldots, 0, 0) \in \mathbb{R}^N,$$
$$\vdots$$
$$e_{n-1} \wedge e_n := (0, 0, 0, \ldots, 0, 1) \in \mathbb{R}^N.$$

From this setting we immediately obtain

Lemma 1.2. *The vectors* $e_k \wedge e_\ell$ *with* $k \neq \ell$ *form a basis of* $\mathbb{R}^N$ *which is orthonormal w.r.t. the Euclidean metric, i.e.*

$$(e_i \wedge e_j) \cdot (e_k \wedge e_\ell) = \begin{cases} 1 & \text{if } i = k, \ j = \ell, \ i \neq j, \ k \neq \ell \text{ etc.} \\ 0 & \text{else} \end{cases}.$$

Lemma 1.3. *For two vectors* $X = (x^1, \ldots, x^n)$ *and* $Y = (y^1, \ldots, y^n)$ *it holds*

$$X \wedge Y = \sum_{1 \leq i < j \leq n} (x^i y^j - x^j y^i) e_i \wedge e_j.$$

Proof. We compute

$$X \wedge Y = \left(\sum_{i=1}^n x^i e_i \right) \wedge \left(\sum_{j=1}^n y^j e_j \right) = \sum_{i,j=1}^n x^i y^j e_i \wedge e_j = \sum_{1 \leq i < j \leq n} (x^i y^j - x^j y^i) e_i \wedge e_j,$$

proving the statement.   □

Let us consider an example: For $n = 3$ we have

$$e_1 \wedge e_2 = (1, 0, 0), \quad e_1 \wedge e_3 = (0, 1, 0), \quad e_2 \wedge e_3 = (0, 0, 1),$$

and for two vectors $X = (x^1, x^2, x^3)$ and $Y = (y^1, y^2, y^3)$ we compute

$$X \wedge Y = x^1 y^2 e_1 \wedge e_2 - x^1 y^3 e_1 \wedge e_3 + x^2 y^1 e_2 \wedge e_1 + x^2 y^3 e_2 \wedge e_3$$
$$-x^3 y^1 e_3 \wedge e_1 + x^3 y^2 e_3 \wedge e_2$$
$$= (x^1 y^2 - x^2 y^1) e_1 \wedge e_2 + (x^3 y^1 - x^1 y^3) e_1 \wedge e_3 + (x^2 y^3 - x^3 y^2) e_2 \wedge e_3$$
$$= (x^1 y^2 - x^2 y^1, x^3 y^1 - x^1 y^3, x^2 y^3 - x^3 y^2).$$

Note that *the usual vector product $X \times Y$ in $\mathbb{R}^3$ does not coincide with the exterior product $X \wedge Y$*, since

$$X \times Y = (x^2 y^3 - x^3 y^2, x^3 y^1 - x^1 y^3, x^1 y^2 - x^2 y^1) \neq X \wedge Y.$$

Without proof we want to collect some algebraic and analytical properties of the exterior product.

Lemma 1.4. *For arbitrary vectors $A, B, C \in \mathbb{R}^n$ there hold*

- $(\lambda A) \wedge B = \lambda (A \wedge B)$;
- $(A + B) \wedge C = A \wedge C + B \wedge C$;
- $(A \wedge B)_{u^i} = A_{u^i} \wedge B + A \wedge B_{u^i}$.

Finally, let $X = (x^1, x^2, 0, \ldots, 0)$ and $Y = (y^1, y^2, 0, \ldots, 0)$ such that

$$X, Y \in \mathrm{span}\{e_1, e_2\}.$$

Then it holds

- $X \wedge Y \perp \mathrm{span}\{X \wedge e_3, \ldots, X \wedge e_n, Y \wedge e_3, \ldots, Y \wedge e_n, e_3 \wedge e_n, \ldots, e_{n-1} \wedge e_n\}$.

1.6.7 The Curvature Vector of the Normal Bundle

Now let us come back to the transformation rule for the matrix $\widetilde{\mathbf{S}}_{12}$ from Sect. 1.6.5 which turns out to be the basis for the definition of the following geometric curvature quantity.

Definition 1.14. The *curvature vector of the normal bundle* is given by

$$\mathfrak{S} := \frac{1}{W} \sum_{1 \leq \sigma < \vartheta \leq n} S_{\sigma,12}^{\vartheta} N_\sigma \wedge N_\vartheta .$$

Here $\wedge$ denotes the exterior product between two vectors in $\mathbb{R}^{n+2}$ from the previous paragraph. If $n = 2$ then $\mathfrak{S}$ can be identified with the scalar normal curvature S.

Proposition 1.13. *The curvature vector of the normal bundle neither depends on the parametrization nor on the choice of the ONF. In particular, its length*

$$|\mathfrak{S}| = \sqrt{\mathfrak{S} \cdot \mathfrak{S}} = \sqrt{\frac{1}{W^2} \sum_{1 \le \sigma < \vartheta \le n} (S^{\vartheta}_{\sigma,12})^2}$$

represents a geometric quantity, the so-called curvature of the normal bundle. □

Proof. We check the invariance w.r.t. rotations:
Using our transformation rule from Sect. 1.6.5 we get

$$\sum_{\sigma,\vartheta=1}^{n} \widetilde{S}^{\vartheta}_{\sigma,12} \, \widetilde{N}_{\sigma} \wedge \widetilde{N}_{\vartheta} = \sum_{\sigma,\vartheta=1}^{n} \sum_{\alpha,\beta=1}^{n} \widetilde{S}^{\vartheta}_{\sigma,12} (R^{\alpha}_{\sigma} N_{\alpha}) \wedge (R^{\beta}_{\vartheta} N_{\beta})$$

$$= \sum_{\sigma,\vartheta=1}^{n} \sum_{\alpha,\beta=1}^{n} (R^{\sigma}_{\alpha})^t \widetilde{S}^{\vartheta}_{\sigma,12} R^{\beta}_{\vartheta} \, N_{\alpha} \wedge N_{\beta}$$

$$= \sum_{\alpha,\beta=1}^{n} S^{\beta}_{\alpha,12} \, N_{\alpha} \wedge N_{\beta}$$

which already proves the statement. □

We want to point out that due to our definition we can distinguish positive and negative signs of the normal curvature S in the case $n = 2$. In contrast to this special situation, the normal curvature is vector-valued if $n > 2$, so that in general we can not speak of "negatively" or "positively" curved normal bundles.

It seems that from the point of view of geometric analysis, *immersions with prescribed normal curvature vector* $\mathfrak{S}$ have not been considered so far. For example, as far as I am aware, results concerning curvature estimates and theorems of Bernstein-type for such special surfaces, comparable e.g. with Bergner and Fröhlich [8], Jost and Xin [76], Wang [119] or Xin [127], do not exist. For a short discussion on this subject we refer to the next chapter.

1.6.8 The Hopf Vector

Let an immersion X in $\mathbb{R}^3$ with second fundamental form L_{ij} be given. In 1950, Hopf [72] showed that the following complex-valued function

$$L_{11}(w) - L_{22}(w) - 2i L_{12}(w)$$

is holomorphic if the scalar mean curvature H of X is constant.

As an application of the integrability conditions of Codazzi and Mainardi we want to conclude this first chapter with a generalization of Hopf's result for minimal surfaces of arbitrary codimension. In particular, this will allow us to characterize the zeros of their Gaussian curvature.

We start with

Proposition 1.14. *Let the conformally parametrized minimal surface X together with an ONF $\mathfrak{N}$ be given. Then its complex-valued Hopf vector*

$$\mathcal{H} = (\mathcal{H}_1, \ldots, \mathcal{H}_n) \in \mathbb{C}^n \quad \text{with} \quad \mathcal{H}_\sigma := L_{\sigma,11} - L_{\sigma,22} - 2i L_{\sigma,12}$$

satisfies the first-order Pascali system

$$\mathcal{H}_{\overline{w}}^t := \frac{1}{2}(\partial_u + i \partial_v)\mathcal{H}^t = \frac{1}{2} \mathbf{T} \circ \mathcal{H}^t$$

with the complex-valued matrix $\mathbf{T} = (T_{\sigma,1}^\vartheta + i T_{\sigma,2}^\vartheta)_{\sigma,\vartheta=1,\ldots,n} \in \mathbb{C}^{n \times n}$.

The proof of this proposition makes use of the following

Lemma 1.5. *Let the conformally parametrized minimal surface X together with an ONF $\mathfrak{N}$ be given. Then there hold*

$$\partial_{\overline{w}}\mathcal{H}_\sigma = \sum_{\vartheta=1}^n \left\{ (-L_{\vartheta,11} + i L_{\vartheta,12})T_{\vartheta,1}^\sigma - (L_{\vartheta,12} + i L_{\vartheta,11})T_{\vartheta,2}^\sigma \right\}$$

for all $\sigma = 1, \ldots, n$.

Proof. Consider the auxiliary functions

$$\mathcal{H}_\sigma^* := L_{\sigma,11} - i L_{\sigma,12}, \quad \sigma = 1, \ldots, n,$$

satisfying $2\mathcal{H}_\sigma^* = \mathcal{H}_\sigma$ due to $L_{\sigma,11} = -L_{\sigma,22}$ for all $\sigma = 1, \ldots, n$. From the Codazzi–Mainardi integrability conditions from Sect. 1.5.2 we infer

$$\partial_v L_{\sigma,11} - \partial_u L_{\sigma,12} = \sum_{\omega=1}^n L_{\omega,12}T_{\omega,1}^\sigma - \sum_{\omega=1}^n L_{\omega,11}T_{\omega,2}^\sigma ,$$

$$\partial_u L_{\sigma,11} + \partial_v L_{\sigma,12} = -\sum_{\omega=1}^n L_{\omega,11}T_{\omega,1}^\sigma - \sum_{\omega=1}^n L_{\omega,12}T_{\omega,2}^\sigma .$$

using conformal parameters together with the conformal representation of the Christoffel symbols from Sect. 1.4.2 Therefore,

$$\partial_{\overline{w}}\mathcal{H}_\sigma^* = \frac{1}{2}(\partial_u L_{\sigma,11} + \partial_v L_{\sigma,12}) + \frac{i}{2}(\partial_v L_{\sigma,11} - \partial_u L_{\sigma,12})$$

$$= -\frac{1}{2}\sum_{\omega=1}^n \left(L_{\omega,11}T_{\omega,1}^\sigma + L_{\omega,12}T_{\omega,2}^\sigma \right) + \frac{i}{2}\sum_{\omega=1}^n \left(L_{\omega,12}T_{\omega,1}^\sigma - L_{\omega,11}T_{\omega,2}^\sigma \right).$$

Rearranging proves the statement.   □

Proof of the proposition. First we note that

$$(-L_{\vartheta,11} + iL_{\vartheta,12})T^\sigma_{\vartheta,1} - (L_{\vartheta,12} + iL_{\vartheta,11})T^\sigma_{\vartheta,2}$$
$$= -(L_{\vartheta,11} - iL_{\vartheta,12})T^\sigma_{\vartheta,1} - i(L_{\vartheta,11} - iL_{\vartheta,12})T^\sigma_{\vartheta,2}$$
$$= -\mathscr{H}^*_\vartheta(T^\sigma_{\vartheta,1} + iT^\sigma_{\vartheta,2}) = \mathscr{H}^*_\vartheta(T^\vartheta_{\sigma,1} + iT^\vartheta_{\sigma,2})$$

which implies

$$\partial_{\overline{w}}\mathscr{H}_\sigma = \frac{1}{2}\sum_{\vartheta=1}^{n}(T^\vartheta_{\sigma,1} + iT^\vartheta_{\sigma,2})\mathscr{H}_\vartheta \quad \text{or} \quad \partial_{\overline{w}}\mathscr{H}^t = \frac{1}{2}\,\mathbf{T}\circ\mathscr{H}^t\,.$$

This proves the proposition.   □

Consequently, we can apply the *local similarity principle for generalized analytic vector-valued functions* from Wendland [121], Theorem 5.3.3, Buchanan [17] or Buchanan and Gilbert [18], Chap. 3, which states that the Hopf vector $\mathscr{H}$ can be represented in the form

$$\mathbf{M}\circ\Phi$$

in a sufficiently small neighborhood $\Omega \subset \mathring{B}$ of a point $w_0 \in \mathring{B}$, with a non-singular and continuous matrix $\mathbf{M} \in \mathbb{C}^{n\times n}$ satisfying $\det \mathbf{M} \neq 0$, and a holomorphic column vector $\Phi \in \mathbb{C}^n$. Furthermore, if $\mathscr{H} \not\equiv 0$ then $\mathscr{H}$ has there at most isolated zeros of finite order.[8]

The point is that the zeros of the Hopf vector $\mathscr{H}$ agree with the zeros of the Gaussian curvature K of the minimal surface X. To see this we compute

$$|\mathscr{H}_\sigma|^2 = (L_{\sigma,11} - L_{\sigma,22})^2 + 4L^2_{\sigma,12} = (L_{\sigma,11} + L_{\sigma,22})^2 - 4(L_{\sigma,11}L_{\sigma,22} - L^2_{\sigma,12})$$
$$= 4(H^2_\sigma - K_\sigma)W^2 = -4K_\sigma W^2$$

for all $\sigma = 1,\ldots,n$. Therefore,

$$|\mathscr{H}|^2 = \sum_{\sigma=1}^{n}|\mathscr{H}_\sigma|^2 = -4\sum_{\sigma=1}^{n}K_\sigma W^2 = -4KW^2\,.$$

Corollary 1.1. *Either the Gaussian curvature K of the minimal immersion X vanishes identically and the surface is a plane, or there exist at most isolated zeros of K in every compact subset $\Theta \subset \mathring{B}$.*

In case $n = 1$ of one codimension, the Hopf function $\mathscr{H} \in \mathbb{C}$ turns out to be holomorphic as a solution of the classical Cauchy–Riemann equation since there are no torsion coefficients.

---

[8]For this generalization of Carleman's theorem see Wendland [121], Theorem 5.3.8.

Chapter 2
Elliptic Systems

Abstract This is an intermediate chapter which first introduces into the theory of non-linear elliptic systems with quadratic growth in the gradient, and which presents secondly some results concerning curvature estimates and theorems of Bernstein-type for surfaces in Euclidean spaces of arbitrary dimensions.

A famous result of S. Bernstein states that a smooth minimal graph in $\mathbb{R}^3$, defined on the whole plane $\mathbb{R}^2$, must necessarily be a plane. Today we know various strategies to prove this result, and the idea goes back to E. Heinz to establish first a curvature estimate and to deduce Bernstein's result in a second step. However, minimal surfaces with higher codimensions do not share this Bernstein property, as one of our main examples $X(w) = (w, w^2) \in \mathbb{R}^4$ with $w = u + iv$ convincingly shows. It is still a great challenge to find geometrical criteria, preferably in terms of the curvature quantities of the surfaces' normal bundles, which guarantee the validity of Bernstein's theorem.

We must admit that we can only discuss briefly some points where we would wish to employ our tools we develop in this book, but up to now we can not continue to drive further developments.

2.1 The Mean Curvature Vector

2.1.1 Mean Curvature and Mean Curvature Vector

Elliptic systems with quadratic growth in the gradient of the form

$$|\Delta Z| \leq a_0 |\nabla Z|^2$$

with the Euclidean Laplace operator Δ and the Euclidean gradient ∇ will play an important role in our analysis. It particularly turns out that the Euler–Lagrange equations for normal Coulomb frames satisfy such non-linear elliptic systems.

S. Fröhlich, *Coulomb Frames in the Normal Bundle of Surfaces in Euclidean Spaces*,
Lecture Notes in Mathematics 2053, DOI 10.1007/978-3-642-29846-2_2,
© Springer-Verlag Berlin Heidelberg 2012

The construction of normal Coulomb frames thus requires a profound knowledge of analytical properties of the underlying geometrical objects. For this reason we devote this intermediate chapter to present some basic facts of conformally parametrized immersions with prescribed mean curvature vector $\mathfrak{H}$ as the standard example of a non-linear elliptic system of the type from above.

Definition 2.1. Let the immersion X together with an ONF $\mathfrak{N}$ be given. Then the *mean curvature* H_{N_σ} of an immersion X *w.r.t. an unit normal vector* $N_\sigma \in \mathfrak{N}$ is defined as

$$H_{N_\sigma} := \frac{1}{2} \sum_{i,j=1}^{2} g^{ij} L_{N_\sigma,ij} = \frac{L_{N_\sigma,11} g_{22} - 2 L_{N_\sigma,12} g_{12} + L_{N_\sigma,22} g_{11}}{2W^2}.$$

Consider an ONF $\mathfrak{N} = (N_1, \ldots, N_n)$, and set $H_\sigma := H_{N_\sigma}$ for abbreviation.

Definition 2.2. The *mean curvature vector* $\mathfrak{H} \in \mathbb{R}^n$ of the immersion X is given by

$$\mathfrak{H} := \sum_{\sigma=1}^{n} H_\sigma N_\sigma.$$

For surfaces in $\mathbb{R}^3$ there is, up to orientation, exactly one unit normal vector N and thus exactly one *mean curvature*

$$H = \frac{L_{11} g_{22} - 2 L_{12} g_{12} + L_{22} g_{11}}{2W^2}.$$

Nevertheless, sometimes ones speaks of the mean curvature vector $\mathfrak{H} = HN$ even in this case of one codimension.

It misleads to believe that the mean curvature vector $\mathfrak{H}$ could *replace* this special unit normal vector N for surfaces in $\mathbb{R}^3$. This is not the case since, for example, for minimal surfaces it always holds $\mathfrak{H} \equiv 0$ while, of course, N does not vanish.

Definition 2.3. The immersion X is called a *minimal surface* if and only if

$$\mathfrak{H} \equiv 0 \quad \text{in } \overline{B}.$$

The property $\mathfrak{H} \equiv 0$ does neither depend on the choice of the normal frame $\mathfrak{N}$ nor on the choice of the parametrization.

In fact, in general it holds: *The mean curvature vector $\mathfrak{H}$ neither depends on the parametrization (if we only admit regular parameter transformations which do not affect the orientation of the unit normal vectors $N_\sigma \in \mathfrak{N}$) nor on the choice of the ONF (if we only admit transformations of class $SO(n)$ between those frames).*

Minimal surfaces are the topic of a huge amount of literature: Courant [28], Nitsche [92], Osserman [94], Dierkes et al. [34], Colding and Minicozzi [27], Eschenburg and Jost [44] to enumerate only some few significant contributions and to illustrate the importance of this surface class in the fields of geometric analysis.

2.1.2 Parallel Mean Curvature Vector

Surfaces with constant mean curvature vector generalize the minimal surface concept. They actually play a central role in modern geometric analysis of surfaces with one codimension $n = 1$ which are immersed in Riemannian or Lorentzian spaces. We refer the reader e.g. to the classical textbook Kenmotsu [78], or to the extensive works of Große-Brauckmann, Heinz, Hildebrandt, Karcher, Korevaar, Kusner, Lawson, Meeks, Sauvigny, Sullivan, Wente, and many others; see for example [54] and the references therein.

So assume now that the mean curvature vector of the immersion X satisfies

$$|\mathfrak{H}| \equiv \text{const} \quad \text{in } \overline{B},$$

where, of course, $\mathfrak{H} \equiv 0$ is allowed. Differentiation yields

$$\sum_{\sigma=1}^{n} H_\sigma H_{\sigma,u^i} = 0,$$

and therefore

$$\sum_{\sigma=1}^{n} H_\sigma H_{\sigma,u^i} - \sum_{\sigma=1}^{n} \sum_{\vartheta=1}^{n} H_\sigma H_\vartheta T_{\sigma,i}^{\vartheta} = 0$$

since the double sum is zero due to $T_{\sigma,i}^{\vartheta} = -T_{\vartheta,i}^{\sigma}$.

We want to associate this property with the following concept.

Definition 2.4. The mean curvature vector $\mathfrak{H}$ of the immersion X is called *parallel in the normal bundle* if the normal parts of the partial derivatives $\partial_{u^i}\mathfrak{H}$ vanish identically, i.e. if there hold

$$\partial_{u^i}^{\perp}\mathfrak{H} \equiv 0 \quad \text{in } B \quad \text{for } i = 1,2.$$

For reasons of simplicity we want to concentrate on the case $n = 2$ of two codimensions. Then the following interesting result holds true (see e.g. Chen [21], or Kenmotsu and Zhou [79] and the references therein).

Proposition 2.1. *If the mean curvature vector $\mathfrak{H}$ of the immersion $X : \overline{B} \to \mathbb{R}^4$ is parallel in the normal bundle then it has constant length. If additionally $\mathfrak{H} \neq 0$, then it holds $S \equiv 0$ for the scalar curvature of the normal bundle.*

Proof. The identities

$$\partial_{u^i}^{\perp}\mathfrak{H} = \sum_{\sigma=1}^{n} H_{\sigma,u^i} N_\sigma + \sum_{\vartheta=1}^{n} H_\vartheta N_{\vartheta,u^i}^{\perp} = \sum_{\sigma=1}^{2} H_{\sigma,u^i} N_\sigma + \sum_{\vartheta=1}^{2}\sum_{\sigma=1}^{2} H_\vartheta T_{\vartheta,i}^{\sigma} N_\sigma = 0$$

for $i = 1, 2$ can be written in the form

$$H_{1,u} = H_2 T_{1,1}^2 , \quad H_{1,v} = H_2 T_{1,2}^2 , \quad H_{2,u} = -H_1 T_{1,1}^2 , \quad H_{2,v} = -H_1 T_{1,2}^2 .$$

Thus, we compute

$$\frac{1}{2} \partial_u |\mathfrak{H}|^2 = H_1 H_{1,u} + H_2 H_{2,u} = H_1 H_2 T_{1,1}^2 - H_1 H_2 T_{1,1}^2 = 0,$$

$$\frac{1}{2} \partial_v |\mathfrak{H}|^2 = H_1 H_{1,v} + H_2 H_{2,v} = H_1 H_2 T_{1,2}^2 - H_1 H_2 T_{1,2}^2 = 0$$

and infer $|\mathfrak{H}|^2 \equiv$ const. Moreover, it holds

$$
\begin{aligned}
0 = \partial_{uv} H_1 - \partial_{vu} H_1 &= H_{2,u} T_{1,2}^2 + H_2 \partial_u T_{1,2}^2 - H_{2,v} T_{1,1}^2 - H_2 \partial_v T_{1,1}^2 \\
&= -H_1 T_{1,1}^2 T_{1,2}^2 + H_1 T_{1,1}^2 T_{1,2}^2 + H_2 \partial_u T_{1,2}^2 - H_2 \partial_v T_{1,1}^2 \\
&= -H_2 (\partial_v T_{1,1}^2 - \partial_u T_{1,2}^2) \\
&= -H_2 S W
\end{aligned}
$$

and analogously $0 = -H_1 S W$. Therefore, either X is a minimal immersion with $\mathfrak{H} \equiv 0$, or if not then it is a surface with mean curvature vector of constant length greater than zero and with flat normal bundle. The statement is proved.   □

2.1.3   The Mean Curvature System

From the Gauß equations in connection with the conformal representation of the Christoffel symbols from Sect. 1.4.2 we now derive an elliptic system for conformally parametrized immersions X with prescribed mean curvature vector $\mathfrak{H}$.

Proposition 2.2. *Let the conformally parametrized immersion X of prescribed mean curvature vector $\mathfrak{H}$ together with an ONF $\mathfrak{N}$ be given. Then it holds*

$$\Delta X = 2 \sum_{\vartheta=1}^{n} H_\vartheta W N_\vartheta = 2 \mathfrak{H} W \quad \text{in } B.$$

Proof. From the Gauß equations we infer

$$
\begin{aligned}
\Delta X &= (\Gamma_{11}^1 + \Gamma_{22}^1) X_u + (\Gamma_{11}^2 + \Gamma_{22}^2) X_v + \sum_{\vartheta=1}^{n} (L_{\vartheta,11} + L_{\vartheta,22}) N_\vartheta \\
&= \sum_{\vartheta=1}^{n} (L_{\vartheta,11} + L_{\vartheta,22}) N_\vartheta .
\end{aligned}
$$

Here we take into account that

$$\Gamma_{11}^1 + \Gamma_{22}^1 = \frac{W_u}{2W} - \frac{W_u}{2W} = 0,$$

$$\Gamma_{11}^2 + \Gamma_{22}^2 = -\frac{W_v}{2W} + \frac{W_v}{2W} = 0$$

as well as

$$L_{\vartheta,11} + L_{\vartheta,22} = 2H_\vartheta W$$

from the definition of H_ϑ. The statement follows.   □

This system generalizes the *classical mean curvature system*

$$\Delta X = 2HWN \quad \text{in } B$$

from Hopf [72] in case $n = 1$ of one codimension with the scalar mean curvature $H \in \mathbb{R}$ and the unit normal vector N of the surface X.

In particular, we infer that conformally parametrized minimal surfaces represent harmonic vectors, i.e. it then holds

$$\Delta X = 0 \quad \text{in } B$$

which offers the possibility to apply the powerful tools of complex analysis to the differential geometry of minimal surfaces. We will discuss this fact later.

2.1.4 Quadratic Growth in the Gradient: A Maximum Principle

Now we want to give a geometric application of the classical maximum principle for subharmonic functions. Namely, assume there is an upper bound $|\mathfrak{H}| \le h_0$ in $\overline{B}$ be given such that for the conformally parametrized immersion X it holds

$$|\Delta X| \le 2h_0 W \le h_0 |\nabla X|^2 \quad \text{in } B$$

on account of

$$W = \sqrt{(X_u \cdot X_u)(X_v \cdot X_v) - (X_u \cdot X_v)^2} = \sqrt{(X_u \cdot X_u)^2}$$

$$= |X_u||X_u| \le \frac{1}{2}\left(X_u^2 + X_u^2\right) = \frac{1}{2}\left(X_u^2 + X_v^2\right) = \frac{1}{2}|\nabla X|^2.$$

Thus, the surface vector X is *solution of a non-linear elliptic system with quadratic growth in the gradient*.

Proposition 2.3. *Let* $X : \overline{B} \to \mathbb{R}^{n+2}$ *be a conformally parametrized immersion with prescribed mean curvature vector* $\mathfrak{H}$. *Let* $|\mathfrak{H}| \leq h_0$ *in* $\overline{B}$, *and suppose that*

$$h_0 \sup_{(u,v) \in B} |X(u,v)| \leq 1.$$

Then it holds the geometric maximum principle

$$\max_{(u,v) \in \overline{B}} |X(u,v)|^2 = \max_{(u,v) \in \partial B} |X(u,v)|^2.$$

Proof. We remark that the statement is obviously true without the assumption on the conformal parametrization since introducing a conformal parameter system $(u,v) \in \overline{B}$ does not affect the maximum norm of the representation $X(u,v)$. Nevertheless, using conformal parameters we compute

$$\Delta |X|^2 = 2\big(|\nabla X|^2 + X \cdot \Delta X\big) \geq 2\big(|\nabla X|^2 - h_0 |X| |\nabla X|^2\big)$$
$$= 2|\nabla X|^2 (1 - h_0 |X|) \geq 0.$$

Therefore, the vector $|X(u,v)|^2$ is subharmonic, and the statement follows from the classical maximum principle. $\qquad\qquad\square$

Surfaces X with the property

$$h_0 \sup_{(u,v) \in B} |X(u,v)| \leq 1$$

are also called *small solutions* of the mean curvature system in contrast to *large solutions* which do not necessarily obey the maximum principle. We will encounter this fact later again. Minimal surfaces are always small in this sense.

The method of proof we presented here goes already back to Heinz (see also Sauvigny [107], vol. 2, Chap. XII). For further considerations we refer e.g. to Dierkes [33] and the references therein.

2.2   Curvature Estimates

2.2.1   Problem Statement

With this intermediate chapter we also want to draw the reader's attention to the problem of curvature estimates and Bernstein-type theorems for minimal surfaces in higher-dimensional Euclidean spaces. In particular, we have in mind to confront some of the methods and results from this field of geometric analysis with the concepts of extrinsic differential geometry which we developed in the first chapter.

This plan must be left incomplete due to its complexity. We will therefore concentrate on some "light" versions of curvature estimates and their immediate consequences, and we will only discuss briefly more profound approaches and methods.

2.2.2 Estimate of the $S_{\sigma,12}^{\vartheta}$

Our first observation is based upon the representation formula

$$S_{\sigma,12}^{\omega} = \frac{1}{W}(L_{\sigma,11} - L_{\sigma,22})L_{\omega,12} - \frac{1}{W}(L_{\omega,11} - L_{\omega,22})L_{\sigma,12}$$

of the normal curvature tensor from Sect. 1.6.2. Applying the Cauchy–Schwarz inequality gives us

$$|S_{\sigma,12}^{\omega}| \le \frac{1}{2W}(L_{\sigma,11}^2 + 2L_{\sigma,12}^2 + L_{\sigma,22}^2) + \frac{1}{2W}(L_{\omega,11}^2 + 2L_{\omega,12}^2 + L_{\omega,22}^2).$$

On the other hand we verify

$$2H_{\sigma}^2 - K_{\sigma} = \frac{L_{\sigma,11}^2 + 2L_{\sigma,11}L_{\sigma,22} + L_{\sigma,22}^2}{2W^2} - \frac{L_{\sigma,11}L_{\sigma,22} - L_{\sigma,12}^2}{W^2}$$

$$= \frac{L_{\sigma,11}^2 + 2L_{\sigma,12}^2 + L_{\sigma,22}^2}{2W^2}$$

so that we arrive at the

Proposition 2.4. *Let the immersion X together with an ONF $\mathfrak{N}$ be given. Then the components $S_{\sigma,12}^{\omega}$ of the curvature vector of its normal bundle can be estimated as follows*

$$|S_{\sigma,12}^{\omega}| \le (2H_{\sigma}^2 - K_{\sigma})W + (2H_{\omega}^2 - K_{\omega})W \quad \text{for all } \sigma, \omega = 1, \dots, n.$$

In particular, immersions with the property

$$2H_{\sigma}^2 - K_{\sigma} \equiv 0 \quad \text{for all } \sigma = 1, \dots, n$$

have flat normal bundle: $S_{\sigma,12}^{\omega} = 0$. But, in general, bounds for $|S_{\sigma,12}^{\omega}|$ can only be achieved by establishing bounds for the curvatures and the area element W.

The special case of two codimensions $n = 2$ leads us to

$$|S|W \le (2H_1^2 - K_1)W + (2H_2^2 - K_2)W = 2|\mathfrak{H}|^2 W - KW$$

due to $S = \frac{1}{W} S^2_{1,12}$. Integration then yields the estimate

$$2 \iint\limits_B |\mathfrak{H}|^2 W \, du dv \geq \iint\limits_B |S| W \, du dv + \iint\limits_B K W \, du dv$$

which we will employ at the end of Sect. 2.2.11. Guadalupe and Rodriguez in [55] derive this integral inequality in case of compact surfaces without boundary.

The *Willmore functional*

$$\iint\limits_B |\mathfrak{H}|^2 W \, du dv$$

on the left hand side enjoys a special attention of the geometric analysis due to its complexity of its non-linear, fourth-order Euler–Lagrange equations, but also due to its wide range of applications in mathematical biology, chemistry, or physics, see e.g. the pioneering work of Helfrich [65] who discusses the significant role of higher-order geometric functionals in $\mathbb{R}^3$ of the general form

$$\iint\limits_B \left\{ \alpha + \beta(H - H_0)^2 + \gamma K \right\} W \, du dv$$

in the theory of so-called elastic bilayers, α, β, γ and H_0 being material constants.

We want to refer the reader to the classical monograph [125] for Willmore's own introduction into the fascinating problem of determining immersions which are critical or even minimal for this functional named after him.

In e.g. Palmer [95] and the recent work Dall'Acqua [31] we find uniqueness results for the Willmore problem for special boundary data. Dall'Acqua et al. [32] prove existence and classical regularity of Willmore surfaces of catenoid-type which were observed phenomenologically e.g. by Fröhlich and Große-Brauckmann using Ken Brakke's surface evolver, see [47]. Concerning the general boundary value problem we want to refer to Schätzle's paper [108].

Moreover, Rivière [98, 99] extends techniques and results e.g. from Helein [64] to derive a non-linear differential equation in a divergence-type form for critical points of the Willmore functional—the basis for further existence and regularity investigations.

Some of Helein's results, on the other hand, will play an important role in our considerations in the fourth chapter.

Let us finally remark that the integral over the Gaussian curvature on the right hand side of the above inequality can be expressed by the Gauß–Bonnet formula in terms of the geodesic curvature κ_g of the immersion X along the boundary curve ∂B,

$$\iint\limits_B K W \, du dv = \int\limits_{\partial B} \kappa_g \, ds - 2\pi,$$

see e.g. Blaschke and Leichtweiß [12] for more details on this famous identity connecting analysis, topology and differential geometry.

And the conformally invariant functional

$$\iint_B |S| W \, du \, dv$$

measures the *total normal curvature* of the surface. In Sakamoto [101] we find the probably first investigations on critical points of this functional, and this should open new fields in classical differential geometry.

2.2.3 The Special Case of Holomorphic Minimal Graphs

We want to specify the foregoing estimate

$$|S| W \le 2|\mathfrak{H}|^2 W - K W$$

in case of holomorphic minimal graphs.

Proposition 2.5. *Let the minimal graph* $X(w) = (x, \Phi(w))$ *on* $\overline{B}$ *with a holomorphic function*

$$\Phi(w) = \varphi(w) + i \psi(w)$$

be given. Then it holds

$$S(w) = -K(w) \quad \text{for all } w \in \overline{B}.$$

Proof. Making use of the special ONF (see Sect. 1.2.2)

$$N_1 = \frac{1}{\sqrt{1 + |\nabla \varphi|^2}} (-\varphi_u, -\varphi_v, 1, 0),$$

$$N_2 = \frac{1}{\sqrt{1 + |\nabla \varphi|^2}} (\varphi_v, -\varphi_u, 0, 1)$$

we will compute the Gaussian curvature K and the normal curvature scalar S. Since X is minimal we already know $|S| \le (-K)$, and we will verify $S = -K$.

For this purpose, we first note

$$L_{1,11} = \frac{\varphi_{uu}}{\sqrt{1 + |\nabla \varphi|^2}}, \quad L_{1,12} = \frac{\varphi_{uv}}{\sqrt{1 + |\nabla \varphi|^2}}, \quad L_{1,22} = \frac{\varphi_{vv}}{\sqrt{1 + |\nabla \varphi|^2}}$$

as well as

$$L_{2,11} = -\frac{\varphi_{uv}}{\sqrt{1 + |\nabla\varphi|^2}}, \quad L_{2,12} = \frac{\varphi_{uu}}{\sqrt{1 + |\nabla\varphi|^2}} = -\frac{\varphi_{vv}}{\sqrt{1 + |\nabla\varphi|^2}},$$

$$L_{2,22} = \frac{\varphi_{uv}}{\sqrt{1 + |\nabla\varphi|^2}}$$

what leads us to (recall $W = 1 + |\nabla\varphi|^2$)

$$K_1 = \frac{L_{1,11}L_{1,22} - L_{1,12}^2}{W^2} = \frac{\varphi_{uu}\varphi_{vv} - \varphi_{uv}^2}{(1 + |\nabla\varphi|^2)^3},$$

$$K_2 = \frac{L_{2,11}L_{2,22} - L_{2,12}^2}{W^2} = \frac{-\varphi_{uv}^2 + \varphi_{uu}\varphi_{vv}}{(1 + |\nabla\varphi|^2)^3}.$$

Thus, the Gaussian curvature of the holomorphic graph turns out to be

$$K = 2\frac{\varphi_{uu}\varphi_{vv} - \varphi_{uv}^2}{(1 + |\nabla\varphi|^2)^3}.$$

Now let us come to the calculation of S : We have

$$T_{1,1}^2 = \left[\partial_u(1 + |\nabla\varphi|^2)^{-\frac{1}{2}}\right](-\varphi_u, -\varphi_v, 1, 0) \cdot N_2$$

$$+ \frac{1}{\sqrt{1 + |\nabla\varphi|^2}}(-\varphi_{uu}, -\varphi_{uv}, 0, 0) \cdot N_2$$

$$= \frac{1}{1 + |\nabla\varphi|^2}(-\varphi_{uu}, -\varphi_{uv}, 0, 0) \cdot (\varphi_v, -\varphi_u, 0, 1)$$

$$= \frac{\varphi_u\varphi_{uv} - \varphi_v\varphi_{uu}}{1 + |\nabla\varphi|^2},$$

and analogously

$$T_{1,2}^2 = \frac{\varphi_u\varphi_{vv} - \varphi_v\varphi_{uv}}{1 + |\nabla\varphi|^2}.$$

Compute now the derivatives

$$\partial_u T_{1,2}^2 = \frac{\varphi_{uu}\varphi_{vv} + \varphi_u\varphi_{uvv} - \varphi_{uv}^2 - \varphi_v\varphi_{uuv}}{1 + |\nabla\varphi|^2} - \frac{2(\varphi_u\varphi_{vv} - \varphi_v\varphi_{uv})(\varphi_u\varphi_{uu} + \varphi_v\varphi_{uv})}{(1 + |\nabla\varphi|^2)^2},$$

$$\partial_v T_{1,1}^2 = \frac{\varphi_{uv}^2 + \varphi_u\varphi_{uvv} - \varphi_{vv}\varphi_{uu} - \varphi_v\varphi_{uuv}}{1 + |\nabla\varphi|^2} - \frac{2(\varphi_u\varphi_{uv} - \varphi_v\varphi_{uu})(\varphi_u\varphi_{uv} + \varphi_v\varphi_{vv})}{(1 + |\nabla\varphi|^2)^2}.$$

A final calculation of

$$S = \frac{1}{W} \left(\partial_v T_{1,1}^2 - \partial_u T_{1,2}^2 \right)$$

would then show the stated identity.

$\square$

2.2.4 Minimal Surfaces in $\mathbb{R}^3$

Bernstein in 1914 proved the following result (see [10] and Hopf [70, 71]).

Proposition 2.6. *A minimal graph $X(x, y) = (x, y, \zeta(x, y))$ satisfying the minimal surface equation*

$$(1 + \zeta_y^2)\zeta_{xx} - 2\zeta_x \zeta_y \zeta_{xy} + (1 + \zeta_x^2)\zeta_{yy} = 0,$$

defined on the whole plane $\mathbb{R}^2$ and with continuous partial derivatives of first and second order, is necessarily a plane.

This result characterizes insistently the non-linear character of the minimal surface equation in contrast to its linearization

$$\Delta \zeta = \zeta_{xx} + \zeta_{yy} = 0,$$

the Laplace equation, which actually possesses non-flat solutions over $\mathbb{R}^2$.

Bernstein's proof relies essentially on his

Lemma 2.1. *Let $\zeta = \zeta(x, y)$ be bounded and twice continuously differentiable, and suppose it solves*

$$A\zeta_{xx} + 2B\zeta_{xy} + C\zeta_{yy} = 0$$

with coefficients A, B and C which depend on $(x, y, \zeta, \zeta_x, \zeta_y, \zeta_{xx}, \zeta_{xy}, \zeta_{yy})$ and fulfill $AC - B^2 > 0$. Then it necessarily holds $\zeta \equiv const.$

Bernstein verifies that $u = \arctan \zeta_x$ is a solution of such a differential equation, and the boundedness of u implies his proposition.

While Bernstein's method was topological in its nature, Heinz [61] in 1952 gave a completely new proof of Bernstein's principle for minimal graphs by establishing a curvature estimate first, what requires deep analytical estimates of the derivatives of the conformally parametrized minimal surface vector from above and an estimate for its area element from below.

For a comprehensive presentation of the theory of plane harmonic mappings together with this estimate of the area element we also want to refer to Duren [40]. For complete treatments of the theory of non-linear elliptic systems of second order with quadratic growth in the gradient we refer the reader to Heinz [62], Sauvigny [107], vol. 2, or Schulz [111].

The point we want to stress is that Bernstein's principle fails for minimal graphs with higher codimensions, for (w, w^2) is obviously a counter-example. One should find geometric conditions which make this principle hold again.

2.2.5 How a Curvature Estimate Could Work

Let the minimal graph on the closed disc $\overline{B}_R$ of radius $R > 0$ together with an ONF $\mathfrak{N}$ be given. We introduce conformal parameters and obtain a harmonic vector-valued mapping $X: \overline{B} \to \mathbb{R}^{n+2}$.

The Gaussian curvature $K_\sigma(0,0)$ in the origin $(0,0) \in B$ w.r.t. an arbitrary $N_\sigma \in \mathfrak{N}$ can be estimated by

$$-K_\sigma(0,0) \le \frac{|L_{\sigma,11}(0,0)||L_{\sigma,22}(0,0)| + |L_{\sigma,12}(0,0)|^2}{W(0,0)^2}$$

where in the enumerator

$$|L_{\sigma,ij}(0,0)| \le |N_\sigma(0,0)||X_{u^i u^j}(0,0)| \le |X_{u^i u^j}(0,0)|$$

or

$$|L_{\sigma,ij}(0,0)| \le |N_{\sigma,u^i}(0,0)||X_{u^j}(0,0)|.$$

Thus, the problem we are faced with is to find (a) upper bounds for the second derivatives of X, or for its first derivatives and the first derivatives of N_σ in the origin, and (b) to establish a lower bound for the area element $W(0,0)$.

2.2.6 Estimate of the Area Element from Below: The Heinz Lemma

Let the minimal graph $X(x,y) = (x, y, \zeta_1(x,y), \ldots, \zeta_n(x,y))$ on the closed disc $\overline{B}_R$ of radius $R > 0$ be given. Introduce conformal parameters $(u,v) \in \overline{B}$ such that it holds

$$\Delta X(u,v) = 0 \quad \text{in } B.$$

We now consider the *harmonic plane mapping*

$$f(u,v) := \big(x^1(u,v), x^2(u,v)\big), \quad (u,v) \in \overline{B}.$$

Since $X(x,y)$ is a graph, *this mapping represents the reparametrization of the graph into the new form* $X(u,v)$, and therefore the *scaled plane mapping*

$$F: \overline{B} \longrightarrow \overline{B} \quad \text{via } F(u,v) := \frac{1}{R} f(u,v)$$

can be chosen with the properties (see e.g. Sauvigny [107], vol. 2):

- F is one-to-one and satisfies $F(0,0) = (0,0)$.
- F maps the boundary ∂B positively oriented and topologically onto ∂B.
- $J_F(u,v) > 0$ in B for the Jacobian of F.

The Heinz lemma on harmonic plane mappings with all these properties states now the following universal estimate.

Proposition 2.7. *With the Heinz constant* $C_H = \frac{27}{4\pi^2} \approx 0.6839\ldots$ *it always holds*

$$|F_w(0,0)|^2 + |F_{\overline{w}}(0,0)|^2 \geq C_H .$$

This estimate immediately implies a lower bound for the area element, namely

$$W(0,0) = \frac{1}{2}|\nabla X(0,0)|^2 \geq \frac{1}{2}|\nabla x^1(0,0)|^2 + \frac{1}{2}|\nabla x^2(0,0)|^2$$

$$= R^2|F_w(0,0)|^2 + R^2|F_{\overline{w}}(0,0)|^2 \geq R^2 C_H .$$

Actually, Heinz first proved

$$|F_w(0,0)|^2 + |F_{\overline{w}}(0,0)|^2 \geq 1 - \frac{2\pi}{3} + \frac{4}{\pi} \approx 0.1788\ldots$$

while the sharp form given in the proposition above goes back to Hall [57], see e.g. Duren's monograph [40] for more details.

Thus, for a complete curvature estimate it remains to estimate the derivatives of X and/or the derivatives of the unit normal vectors N_σ.

2.2.7 Minimal Surfaces with Controlled Growth

Let the minimal graph be conformally parametrized via $X \colon \overline{B} \to \mathbb{R}^{n+2}$. We have

$$|L_{\sigma,11}||L_{\sigma,22}| + |L_{\sigma,12}|^2 \leq |X_{uu}||X_{vv}| + |X_{uv}|^2 .$$

Due to $\Delta X = 0$, potential theory yields a universal constant $C_1 \in (0,\infty)$ such that

$$|X_{u^i u^j}(0,0)| \leq C_1 \sup_{(u,v) \in B} |X(u,v)| = C_1 \sup_{(x,y) \in B_R} |X(x,y)|,$$

see e.g. Gilbarg and Trudinger [53], Theorem 4.6. Now we arrive at the following curvature estimate and theorem of Bernstein-type from Fröhlich [48].

Theorem 2.1. *Let there exist a constant* $\Omega \in (0,\infty)$ *such that the minimal graph* $X \colon \overline{B}_R \to \mathbb{R}^{n+2}$ *satisfies the following growth condition*

$$|X(x,y)| \leq \Omega R^\varepsilon$$

with some $\varepsilon \in [0, 2)$. Then it holds the curvature estimate

$$|K_\sigma(0,0)| \leq \frac{2C_1^2 \Omega^2}{C_H^2} \cdot \frac{R^{2\varepsilon}}{R^4}.$$

Thus, if the minimal graph is defined over the whole $\mathbb{R}^2$ then it is a plane.

The last statement in this theorem follows after performing the limit $R \to \infty$.

This result is sharp in the following sense: $X(w) = (w, w^2)$, defined on the whole plane $\mathbb{R}^2$, has quadratic growth, i.e. $\varepsilon = 2$ in the terminology of our theorem, and it is obviously not a plane!

We also want to mention that our theorem generalizes the classical Liouville theorem from complex analysis.

It arises the question whether the critical growth $\varepsilon = 2$ has something to do with the non-vanishing of the scalar curvature of the normal bundle. This question must be left open.

2.2.8 The First and Second Variation of the Area Functional

Now we want to draw the reader's attention to curvature estimates for *stable minimal surfaces*. For this purpose we first consider immersions $X \colon \overline{B} \to \mathbb{R}^{n+2}$ which (a) are critical for the area functional

$$\mathscr{A}[X] := \iint_B W \, du\, dv \quad \text{with } W = \sqrt{g_{11} g_{22} - g_{12}^2},$$

and (b) for which its second variation is always positive.

For the next two results we especially refer to Sauvigny [102].

Proposition 2.8. *The immersion $X \colon \overline{B} \to \mathbb{R}^{n+2}$ is critical for $\mathscr{A}[X]$ if its mean curvature vector vanishes identically, i.e. if*

$$\mathfrak{H} \equiv 0 \quad \text{in } B.$$

In other words, minimal surfaces are stationary for the area functional.

Proposition 2.9. *The second variation of $\mathscr{A}[X]$ w.r.t. an unit normal vector $N_\sigma \in \mathfrak{N}$ for a conformally parametrized minimal immersion $X \colon \overline{B} \to \mathbb{R}^{n+2}$ reads*

$$\delta_\sigma^2 \mathscr{A}[X] = \iint_B \left(|\nabla \varphi|^2 + 2K_\sigma W \varphi^2 \right) du\, dv$$

$$+ \sum_{\vartheta=1}^{n} \iint_B \left\{ (T_{\sigma,1}^\vartheta)^2 + (T_{\sigma,2}^\vartheta)^2 \right\} \varphi^2 \, du\, dv$$

for arbitrary $\varphi \in C_0^\infty(B, \mathbb{R})$.

In case $n = 1$ of one codimension there is only one unit normal vector w.r.t. which we can evaluate the second variation. Thus, we would then arrive at

$$\delta^2 \mathscr{A}[X] = \iint\limits_{B} \left(|\nabla \varphi|^2 + 2K\, W \varphi^2 \right) du\, dv$$

since the integral over the squared torsion coefficients drops out.

2.2.9 Stable Minimal Surfaces

The second variation leads us directly to the

Definition 2.5. The minimal surface X is called *stable* if it holds

$$\delta_N^2 \mathscr{A}[X] \geq 0 \quad \text{for all } \varphi \in C_0^\infty(B, \mathbb{R})$$

and all unit normal vectors N.

For fixed ONF $\mathfrak{N}$ and fixed test function $\varphi \in C_0^\infty(B, \mathbb{R})$ we could sum up all the n stability inequalities $\delta_\sigma^2 \mathscr{A}[X] \geq 0$ for $\sigma = 1, \ldots, n$ to get

$$\iint\limits_{B} |\nabla \varphi|^2 \, du\, dv \geq \frac{2}{n} \iint\limits_{B} (-K) W \varphi^2 \, du\, dv$$

$$-\frac{1}{n} \sum_{\sigma, \omega = 1}^{n} \iint\limits_{B} \left\{ (T_{\sigma,1}^\omega)^2 + (T_{\sigma,2}^\omega)^2 \right\} \varphi^2 \, du\, dv,$$

again for all test functions φ. Note that $\mathfrak{H} \equiv 0$ and $K \leq 0$ for the minimal surface.

It must be remarked that *the right hand side of these inequalities depends on the choice of the ONF $\mathfrak{N}$ while the left hand side does not.* Thus, it arises the question whether there exists an ONF $\mathfrak{N}$ with controlled torsion coefficients such that the difference at the right hand side stays positive for all φ.

In the next two chapters we will construct special Coulomb-gauged ONF's for which we can in fact control the torsion by means of the curvature of the normal bundle (and certain smallness conditions in case $n > 2$).

In particular, we will show that *if the normal bundle is flat then there exist an ONF $\mathfrak{N}$ which is free of torsion,* and then the minimal surface is stable if

$$\iint\limits_{B} |\nabla \varphi|^2 \, du\, dv \geq 2 \iint\limits_{B} (-K_N) W \varphi^2 \, du\, dv$$

for all test functions $\varphi \in C_0^\infty(B, \mathbb{R})$ and all unit normal vectors N. It will turn out that the curvature of the normal bundle acts as a barrier for the existence of orthogonal unit normal frames with vanishing torsion coefficients. So if we set

$$T := \sum_{1 \le \sigma < \omega \le n} \left\{ (T_{\sigma,1}^{\omega})^2 + (T_{\sigma,2}^{\omega})^2 \right\}$$

in the general case $T > 0$, when we can not expect existence of torsion-free ONF's, we obtain from the above stability inequality after partial integration

$$0 \le \iint_B \left\{ |\nabla \varphi|^2 + \frac{2}{n}(K+T)W\varphi^2 \right\} du\, dv = \iint_B \left\{ -\Delta\varphi + \frac{2}{n}(K+T)W\varphi \right\} \varphi\, du\, dv.$$

The Schwarzian eigenvalue problem which arises from here,

$$-\Delta\varphi + \lambda(K+T)\varphi = 0 \quad \text{in } B, \quad \varphi = 0 \quad \text{on } \partial B,$$

was first considered in Barbosa and do Carmo [5], later in Sauvigny [102, 104] in his studies of minimal surfaces with polygonal boundaries, but, however, always without taken the curvature of the normal bundle into particular account.

Thus, also here it remains the question whether we can characterize stability of minimal surfaces in terms of the eigenvalues of that Schwarzian eigenvalue problem, and how these eigenvalues depend on the curvature of the normal bundle.

Sauvigny applied his results to prove uniqueness for minimal surfaces spanning so-called extreme polygonal boundary curves, see e.g. [103]. Moreover, in [106] he establishes compactness and finiteness results for stable and unstable small immersions with constant mean curvature spanning regular, extreme Jordan curves.

Concerning new results on finiteness for minimal surfaces with polygonal boundaries we want to draw the reader's attention to the papers Jakob [73–75]. In this context we would also like to refer to a recent result of Bergner and Jakob [9] on the non-existence of branch points for minimal surfaces in $\mathbb{R}^{n+2}$.

Finally, we want to remark that already Wirtinger in [126] proved the absolutely area minimizing property of holomorphic minimal surfaces w.r.t. compactly supported variations which implies stability in the sense of our definition from the beginning. This minimizing character is also discussed in Eschenburg and Jost [44] by means of modern calibration methods.

2.2.10 Osserman's Curvature Estimate and a Generalization

In 1964, Osserman [93] proved the following

Proposition 2.10. *Assume that at each point of a minimal immersion X in $\mathbb{R}^{n+2}$ all unit normal vectors make an angle of at least $\omega > 0$ with a fixed axis in space. Then for the Gauß curvature $K(P)$ at some point $P = X(w_0)$, $w_0 \in B$, with interior distance $d > 0$ to the boundary, it holds*

$$|K(P)| \leq \frac{1}{d^2} \cdot \frac{16(n+1)}{\sin^4 \omega} \, .$$

In particular, if the surface is defined over the whole $\mathbb{R}^2$, then it is a plane.

A connection of Osserman's ω-condition with the curvature of the normal bundle of complete minimal graphs is not known to us. However, his result is sharp in the sense that the holomorphic graph (w, w^2), $w \in \mathbb{R}^2$, does not obey the ω-condition.

A refinement of Osserman's proof together with applications of potential theoretic methods enabled us in Bergner and Fröhlich [8] to prove the following curvature estimate for graphs with prescribed Hölder continuous mean curvature vector.

Proposition 2.11. *Let the graph*

$$X(x, y) = \big(x, y, \zeta_1(x, y), \dots, \zeta_n(x, y)\big)$$

of prescribed mean curvature vector

$$\mathfrak{H} = \mathfrak{H}(X, Z)$$

be given, $(X, Z) \in \mathbb{R}^{n+2} \times S^{n+1}$ with $S^{n+1} \subset \mathbb{R}^{n+2}$ being the $(n+1)$-dimensional unit sphere. Suppose that $X(u, v)$ represents a conformal reparametrization of this graph. Assume furthermore

1. *The mean curvature vector $\mathfrak{H} = \mathfrak{H}(X, Z)$ satisfies*

$$|\mathfrak{H}(X, Z)| \leq h_0 \quad \text{for all } X \in \mathbb{R}^{n+2} \text{ and } Z \in S^{n+1}$$

   and

$$|\mathfrak{H}(X_1, Z_1) - \mathfrak{H}(X_2, Z_2)| \leq h_1 |X_1 - X_2|^\alpha + h_2 |Z_1 - Z_2|$$

   for all $X_1, X_2 \in \mathbb{R}^{n+2}$ and $Z_1, Z_2 \in S^{n+1}$, with real constants $h_0, h_1, h_2 \in [0, \infty)$ and with some $\alpha \in (0, 1)$.
2. *The surface represents (or contains) a geodesic disc $\mathfrak{B}_r(X_0)$ of radius $r > 0$ and center $X_0 \in \mathbb{R}^{n+2}$ (see the next paragraph for details).*
3. *With a real constant $d_0 > 0$, the area of this geodesic disc $\mathfrak{B}_r(X_0)$ can be estimated by*

$$\mathscr{A}[\mathfrak{B}_r(X_0)] \leq d_0 r^2 \, .$$

4. *At every point $w \in B$, each normal vector of X makes an angle of at least $\omega > 0$ with the x_1-axis.*

Then, for an arbitrarily chosen ONF $\mathfrak{N}$ there exists a constant

$$\Theta = \Theta(h_0 r, h_1 r^{1+\alpha}, h_2 r, d_0, \sin \omega, \alpha) \in (0, \infty)$$

such that it holds the curvature estimate

$$|K_\sigma(0,0)| \le \frac{1}{r^2} \left\{ (h_0 r)^2 + \Theta \right\} \quad \textit{for all } \sigma = 1, \dots, n.$$

In particular, if $\mathfrak{H} \equiv 0$, and therefore $h_0, h_1, h_2 = 0$, and if the minimal graph is defined over the whole plane $\mathbb{R}^2$, then $X(x, y)$ must be affine linear.

A few words to the assumptions in this theorem: Since X is a conformally parametrized graph with prescribed mean curvature vector $\mathfrak{H}$, it is a solution of

$$\Delta X = 2 \sum_{\sigma=1}^{n} H_\sigma W N_\sigma = 2\mathfrak{H}W \quad \text{in } B.$$

Together with the first assumption we arrive at the following non-linear elliptic system with quadratic growth in the gradient (see Sect. 2.1.4)

$$|\Delta X| \le h_0 |\nabla X|^2 \quad \text{in } B.$$

The first and second derivatives (in the interior) of such a system can be controlled if X is either a small solution in the sense of

$$h_0 \cdot \sup_{(u,v) \in B} |X(u,v)| < 1,$$

or if a growth condition for the area as required is known such that a smallness condition can eventually be forced by means of the Courant–Lebesgue lemma in connection with a geometric maximum principle.

On the other hand, the assumption on the universal angle ω as well as the required graph property are needed (a) to ensure that also the plane mapping

$$f(u,v) = \left(x^1(u,v), x^2(u,v) \right), \quad (u,v) \in \overline{B},$$

solves a non-linear elliptic system with quadratic growth in the gradient, and (b) that it is one-to-one with $F(0,0) = (0,0)$, positively oriented and topologically on the boundary, and possesses a positive Jacobian in B, see Sect. 2.2.6.

Most of our inputs were already discussed in the foregoing paragraphs. But also here the question remains open whether these assumptions can be connected to the inner geometry of the normal bundle.

2.2.11 On the Growth of Geodesic Discs

At least we can give a partial answer to this question regarding the growth condition for geodesic discs. Is it valid to require such a condition at all? In Bergner and

Fröhlich [8] we computed directly

$$\mathscr{A}[X] \leq 192\pi r^2$$

for geodesic discs of the holomorphic graph (w, w^2), where $r > 0$ is chosen sufficiently large. Note that in this special case the scalar curvature S of the normal bundle vanishes asymptotically.

If S is otherwise everywhere strictly larger than a positive constant for some immersion $X: \overline{B} \to \mathbb{R}^4$ (or smaller than a negative constant), we can show

Proposition 2.12. *Let a minimal surface $X: \overline{B} \to \mathbb{R}^4$ be given such that the scalar curvature S of its normal bundle satisfies*

$$S(u, v) \geq S_0 > 0 \quad \text{for all } (u, v) \in \overline{B}$$

with a fixed real number $S_0 > 0$.
Suppose furthermore that X represents (or contains) a geodesic disc $\mathfrak{B}_r(X_0)$ with geodesic radius $r > 0$ and with center X_0. Then for the area of this disc it holds the estimate

$$\mathscr{A}[\mathfrak{B}_r(X_0)] \geq \pi r^2 + \frac{S_0 \pi}{12} r^4$$

Proof. Let the geodesic disc $\mathfrak{B}_r(X_0)$ be given parametrically as $X(\rho, \varphi)$ with geodesic polar coordinates $(\rho, \varphi) \in [0, r] \times [0, 2\pi]$. With the area element $\sqrt{P(\rho, \tau)}$, the line element ds_P^2 w.r.t. this coordinate system takes the form

$$ds_P^2 = d\rho^2 + P(\rho, \varphi) \, d\varphi,$$

with smooth $P(\rho, \varphi) > 0$ for all $(\rho, \varphi) \in (0, r] \times [0, 2\pi)$ satisfying

$$\lim_{\rho \to 0+} P(\rho, \varphi) = 0, \quad \lim_{\rho \to 0+} \frac{\partial}{\partial \rho} \sqrt{P(\rho, \varphi)} = 1$$

for all $\varphi \in [0, 2\pi)$. For these results and for the following identities we refer the reader e.g. to Blaschke and Leichtweiß [12]. In particular, with the geodesic curvature κ_g of the surface, the integral formula of Gauß–Bonnet gives

$$\int_0^r \kappa_g(\rho, \varphi) \sqrt{P(\rho, \varphi)} \, d\varphi + \int_0^\rho \int_0^{2\pi} K(\tau, \varphi) \sqrt{P(\tau, \varphi)} \, d\tau d\varphi = 2\pi.$$

For curves with $\rho = \text{const}$ it holds

$$\kappa_g(\rho, \varphi) \sqrt{P(\rho, \varphi)} = \frac{\partial}{\partial \rho} \sqrt{P(\rho, \varphi)} \quad \text{for all } (\rho, \varphi) \in (0, r] \times [0, 2\pi),$$

and therefore, together with $-K \geq |S| \geq S_0 > 0$ (see Sect. 2.2.2) we can estimate

$$
\frac{\partial}{\partial \rho} \int_0^{2\pi} \sqrt{P(\rho, \varphi)} \, d\varphi = \int_0^{2\pi} \kappa_g(\rho, \varphi) \sqrt{P(\rho, \varphi)} \, d\varphi
$$

$$
= 2\pi - \int_0^{\rho} \int_0^{2\pi} K(\tau, \varphi) \sqrt{P(\tau, \varphi)} \, d\tau d\varphi
$$

$$
\geq 2\pi + \int_0^{\rho} \int_0^{2\pi} S_0 \sqrt{P(\tau, \varphi)} \, d\tau d\varphi
$$

$$
= 2\pi + S_0 \mathscr{A}[\mathfrak{B}_\rho(X_0)]
$$

$$
\geq 2\pi + S_0 \pi \rho^2
$$

for all $\rho \in (0, r]$. Integration over the radius coordinate yields

$$
\int_0^{2\pi} \sqrt{P(\rho, \varphi)} \, d\varphi \geq 2\pi \rho + \frac{S_0 \pi}{3} \rho^3,
$$

and a further integration over $\rho = 0 \dots r$ shows

$$
\mathscr{A}[\mathfrak{B}_r(X_0)] = \int_0^{r} \int_0^{2\pi} \sqrt{P(\rho, \varphi)} \, d\varphi d\rho \geq \pi r^2 + \frac{S_0 \pi}{12} r^4
$$

proving the statement. $\qquad\qquad\qquad\qquad\qquad\qquad\qquad\qquad\qquad\qquad$ □

At least for certain stable minimal geodesic discs we can show that their areas grow quadratically in the radius r. Namely, let us start again from the stability inequality

$$
\delta_\omega^2 \mathscr{A}[X] = \iint_B |\nabla \varphi|^2 \, du dv + 2 \iint_B K_\omega W \varphi^2 \, du dv
$$

$$
+ \sum_{\sigma=1}^{n} \iint_B \left\{ (T_{\omega,1}^\sigma)^2 + (T_{\omega,2}^\sigma)^2 \right\} \varphi \, du dv
$$

$$
\geq 0
$$

for all $\varphi \in C_0^\infty(B, \mathbb{R})$ and using conformal parameters. Since for the non-positive Gauß curvature K we know

$$
K_\omega \geq \sum_{\sigma=1}^{n} K_\sigma = K,
$$

we have the estimate

$$\delta^2_\omega \mathscr{A}[X] \geq D^2 \mathscr{A}[X] := \iint_B |\nabla \varphi|^2 \, du \, dv + 2 \iint_B K W \varphi^2 \, du \, dv$$

for all $\varphi \in C_0^\infty(B, \mathbb{R})$. The integral $D^2 \mathscr{A}[X]$ agrees with the functional of the second variation for minimal surfaces in $\mathbb{R}^3$ such that in this situation the condition

$$D^2 \mathscr{A}[X] \geq 0 \quad \text{for all } \varphi \in C_0^\infty(B, \mathbb{R})$$

actually defines *stability for minimal surfaces* $X: \overline{B} \to \mathbb{R}^3$.

Sauvigny in [102] revives this condition to define (strict) stability for minimal immersions $X: \overline{B} \to \mathbb{R}^{n+2}$. It holds the

Proposition 2.13. *Let the geodesic disc* $\mathfrak{B}_r(X_0)$ *be stable in the sense of*

$$D^2 \mathscr{A}[\mathfrak{B}_r(X_0)] \geq 0 \quad \text{for all } \varphi \in C_0^\infty(B, \mathbb{R}).$$

Then its area can be estimates by

$$\mathscr{A}[\mathfrak{B}_r(X_0)] \leq \frac{4\pi}{3} r^2.$$

For the proof we refer the reader to Gulliver [56] and Sauvigny [105].

Two concluding remarks are due: First, and this follows from Sect. 2.2.2, we estimate

$$S_0 \iint_B W \, du \, dv \leq \iint_B |S_0| W \, du \, dv \leq \iint_B (-K) W \, du \, dv,$$

that is, *if* $S_0 \neq 0$ *then the curvatura integra is not finite for complete minimal graphs.*

And secondly, and this is to round out the beginning of this paragraph, Micallef in [88] showed that if a minimal graph is complete and stable, and if its area growths quadratically, then it is holomorphic. Wirtinger in [126] proved that holomorphic minimal surfaces area absolutely area minimizing w.r.t. compactly supported variations, see our discussion in Sect. 2.2.9 above.

2.2.12 Curvature Estimates for Higher-Dimensional Minimal Graphs

The next result, which goes back to Hildebrandt, Jost and Widman [68], states a Bernstein-type result for higher-dimensional minimal surface graphs in Euclidean spaces of arbitrary dimensions.

In particular, it contains a gradient bound of the graph a priori which essentially reflects the fact that a generalized Gauß map of the surface (concerning this, see also Hoffman and Osserman [69]) must be contained in a geodesic ball of radius $\frac{\sqrt{2}\pi}{4}$ in the so-called Grassmannian manifold $G_{r,s}$.

Proposition 2.14. *Let $z^\alpha = f^\alpha(x)$ with $\alpha = 1, \ldots, s$ and $x = (x^1, \ldots, x^r) \in \mathbb{R}^r$ be C^2-regular, and let it generate a r-dimensional minimal graph*

$$X(x,y) = \left(x^1, \ldots, x^r, f^1(x^1, \ldots, x^r), \ldots, f^s(x^1, \ldots, x^r)\right).$$

Let there furthermore exist a real number $\beta > 0$ such that

$$\beta < \cos^{-t}\left(\frac{\pi}{2\sqrt{t}\,K}\right)$$

with

$$K := \begin{cases} 1 & \text{if } t = 1 \\ 2 & \text{if } t \geq 2 \end{cases} \quad \text{and} \quad t := \min\{r, s\}.$$

Assume finally that

$$\sqrt{\det\left(\delta_{ij} + \sum_{\alpha=1}^{s} \frac{\partial f^\alpha}{\partial x^i} \frac{\partial f^\alpha}{\partial x^j}\right)_{i,j=1,\ldots,r}} < \beta.$$

Then the functions $f^1, \ldots, f^s$ are affine linear on $\mathbb{R}^r$, and the minimal graph is an affine linear r-dimensional plane.

In subsequent works, e.g. Jost and Xin [76], Wang [119], or Xin [127], one finds various improvements of this result as well as generalizations to surfaces with prescribed mean curvature vector.

Also here the question must be left open how the assumption on the gradient bound stands in connection with our geometrical and analytical concepts of the normal bundle. We expect that it can be weakened at least in the special case $r = 2$ of surfaces in view of other results where less restrictive assumptions are required, see e.g. Fröhlich [46] and the references therein, and combine them for instance with the discussions from Barbosa and do Carmo [5, 6], and Ruchert [100].

Finally we want to mention the curvature estimates and theorems of Bernstein-type for minimal submanifolds with flat normal bundle from Smoczyk et al. [113], where e.g. classical methods from Schoen et al. [109] and Ecker, Huisken [42] were employed. In Fröhlich and Winklmann [52] we succeeded in proving similar results for graphs of dimension $m \in [2, 5]$ but with prescribed mean curvature vector. Can one find methods and techniques comparable to the ones presented in this book to establish more general results for submanifolds with *arbitrary normal bundles?*

Chapter 3
Normal Coulomb Frames in $\mathbb{R}^4$

Abstract With this chapter we begin our study of constructing normal Coulomb frames, here for surfaces immersed in Euclidean space $\mathbb{R}^4$.

Normal Coulomb frames are critical for a new functional of total torsion. We present the associated Euler–Lagrange equation and discuss its solution via a Neumann boundary value problem. A proof of the "minimal character" of normal Coulomb frames follows immediately.

Using methods from potential theory and complex analysis we establish various analytical tools to control these special frames. For example, we present two different methods to bound their torsion (connection) coefficients. Methods from the theory of generalized analytic functions will play again an important role.

We conclude the third chapter with a class of minimal graphs for which we can explicitly compute normal Coulomb frames.

3.1 Torsion-Free Normal Frames for Curves and Surfaces

3.1.1 Curves in $\mathbb{R}^3$

In this third chapter we want to work out methods for a new analytical description of the normal bundles of surfaces in $\mathbb{R}^4$. This particularly contains the construction of so-called *normal Coulomb frames* together with a comprehensive discussion of their regularity properties.

To begin with we consider a regular arc-length parametrized curve $c(s)$ in $\mathbb{R}^3$ with

$$\text{unit tangent vector } \mathfrak{t}(s) = c'(s),$$

$$\text{unit normal vector } \mathfrak{n}(s) = \frac{\mathfrak{t}'(s)}{|\mathfrak{t}'(s)|} \text{ and}$$

$$\text{unit binormal vector } \mathfrak{b}(s) = \mathfrak{t}(s) \times \mathfrak{n}(s).$$

S. Fröhlich, *Coulomb Frames in the Normal Bundle of Surfaces in Euclidean Spaces*, 53
Lecture Notes in Mathematics 2053, DOI 10.1007/978-3-642-29846-2_3,
© Springer-Verlag Berlin Heidelberg 2012

The *torsion* $\tau(s)$ of this curve is then given by

$$\tau(s) = \mathfrak{n}(s)' \cdot \mathfrak{b}(s) = -\mathfrak{n}(s) \cdot \mathfrak{b}(s)' \, .$$

Now we introduce a *new unit normal frame* $(\widetilde{\mathfrak{n}}, \widetilde{\mathfrak{b}})$ by means of a $SO(2)$-action, i.e. by a usual rotation, as follows

$$\widetilde{\mathfrak{n}} = \cos \varphi \, \mathfrak{n} + \sin \varphi \, \mathfrak{b}, \quad \widetilde{\mathfrak{b}} = -\sin \varphi \, \mathfrak{n} + \cos \varphi \, \mathfrak{b}$$

with an angle of rotation $\varphi = \varphi(s)$. The *new torsion* $\widetilde{\tau}$ associated to this new frame then results from the following computation

$$\begin{aligned}
\widetilde{\tau} = \widetilde{\mathfrak{n}}' \cdot \widetilde{\mathfrak{b}} &= (-\varphi' \sin \varphi \, \mathfrak{n} + \cos \varphi \, \mathfrak{n}' + \varphi' \cos \varphi \, \mathfrak{b} + \sin \varphi \, \mathfrak{b}') \cdot (-\sin \varphi \, \mathfrak{n} + \cos \varphi \, \mathfrak{b}) \\
&= \varphi' \sin^2 \varphi - \sin^2 \varphi \, (\mathfrak{b}' \cdot \mathfrak{n}) + \cos^2 \varphi \, (\mathfrak{n}' \cdot \mathfrak{b}) + \varphi' \cos^2 \varphi \\
&= \varphi' + \tau.
\end{aligned}$$

In particular, constructing an ONF $(\widetilde{\mathfrak{n}}, \widetilde{\mathfrak{b}})$ which is *free of torsion*, i.e. which fulfills $\widetilde{\tau} \equiv 0$, by starting with a given ONF $(\mathfrak{n}, \mathfrak{b})$, reduces to solving the problem

$$\varphi'(s) = -\tau(s), \quad \varphi(s_0) = \tau_0$$

with some initial value τ_0.

Proposition 3.1. *Rotating an initial ONF* $(\mathfrak{n}, \mathfrak{b})$ *by an angle*

$$\varphi(s) = -\int_{s_0}^{s} \tau(\sigma) \, d\sigma + \varphi_0$$

with arbitrary $\varphi_0 \in \mathbb{R}$ *generates an ONF* $(\widetilde{\mathfrak{n}}, \widetilde{\mathfrak{b}})$ *which is free of torsion.*

Such a torsion-free ONF is also called *parallel* because *all of its derivatives are tangential to the curve,* that is *parallel to* $\mathfrak{t}(s)$. Moreover, *they are special normal Coulomb frames* what will become clear in the following.

Parallel ONF's for curves are widely used in geometry and mathematical physics. We want to refer again to Burchard and Thomas [19] and the references therein for an application of such frames in the theory of Euler's elastic curves, see our short discussion in Sect. 1.3.5.

3.1.2   Torsion-Free Normal Frames

The question now arises whether there is a similar construction of torsion-free ONF's *for two-dimensional surfaces.*

In this chapter we focus on the case of two codimensions $n = 2$. So let us given an ONF $\mathfrak{N} = (N_1, N_2)$. By means of the $SO(2)$-valued transformation

$$\widetilde{N}_1 = \cos\varphi \, N_1 + \sin\varphi \, N_2, \quad \widetilde{N}_2 = -\sin\varphi \, N_1 + \cos\varphi \, N_2$$

with a rotation angle φ we get a new ONF $\widetilde{\mathfrak{N}} = (\widetilde{N}_1, \widetilde{N}_2)$.

Lemma 3.1. *Let* $X : \overline{B} \to \mathbb{R}^4$ *be given. Then there hold the transformation formulas*

$$\widetilde{T}^2_{1,1} = T^2_{1,1} + \varphi_u, \quad \widetilde{T}^2_{1,2} = T^2_{1,2} + \varphi_v \quad in \ B$$

for the torsion coefficients $T^\vartheta_{\sigma,i}$ *and* $\widetilde{T}^\vartheta_{\sigma,i}$ *of two ONF's* $\mathfrak{N}$ *resp.* $\widetilde{\mathfrak{N}}$.

We omit the proof of this lemma which follows the same lines as our calculation from the beginning of this chapter. Rather we want to compute an angle φ which carries $\mathfrak{N}$ into a new ONF $\widetilde{\mathfrak{N}}$ which is *free of torsion,* i.e. which satisfies

$$\widetilde{T}^2_{1,1} = 0 \quad \text{and} \quad \widetilde{T}^2_{1,2} = 0 \quad \text{everywhere in } \overline{B}.$$

Obviously, φ then has to solve the linear system

$$\varphi_u = -T^2_{1,1} \quad \text{and} \quad \varphi_v = -T^2_{1,2}.$$

Recall that such a system is solvable if and only if the integrability condition

$$0 = -\partial_v\varphi_u + \partial_u\varphi_v = \partial_v T^2_{1,1} - \partial_u T^2_{1,2} = \mathrm{div}\,(-T^2_{1,2}, T^2_{1,1}) = SW \quad \text{in } B$$

is satisfied with the scalar curvature S of the normal bundle from Sect. 1.6.3 and the area element W of the immersion. Since there always exists an ONF $\mathfrak{N}$ thanks to the special topology of the domain of definition $\overline{B}$, we have

Theorem 3.1. *The immersion* $X : \overline{B} \to \mathbb{R}^4$ *admits a torsion-free ONF* $\mathfrak{N}$ *if and only if the scalar curvature* S *of its normal bundle vanishes identically in* $\overline{B}$.

Such a torsion-free ONF *is again parallel in the sense that its derivatives have no normal parts,* for the Weingarten equations from Sect. 1.4.5 now take the form

$$N_{\sigma,u} = -\frac{L_{\sigma,11}}{W} X_u - \frac{L_{\sigma,12}}{W} X_v, \quad N_{\sigma,v} = -\frac{L_{\sigma,12}}{W} X_u - \frac{L_{\sigma,22}}{W} X_v$$

for $\sigma = 1, 2$ and using conformal parameters.

Note also that such an ONF *is not uniquely determined,* rather we can rotate the whole frame by a constant angle φ_0 without effecting the torsion coefficients, because the above differential equations contain only derivatives of φ.

We want to remark that existence of parallel frames in case of vanishing curvature $S \equiv 0$ is in fact well settled. With our next considerations we want to establish existence and regularity of ONF's *if the normal bundle is curved,* and such new frames will replace the concept of parallel frames in this more general situation.

3.1.3 Examples

But first we want to discuss some elementary examples of surfaces with flat normal bundle.[1]

1. Let us begin with *spherical surfaces* characterized by the property

$$|X(u,v)| = 1 \quad \text{for all } (u,v) \in \overline{B}.$$

We immediately compute

$$X_u \cdot X = 0, \quad X_v \cdot X = 0,$$

i.e. X itself is our first unit normal vector, say $X = N_1$. A second one follows after completion of $\{X_u, X_v, N_1\}$ to a basis of the whole embedding space $\mathbb{R}^4$.
  Then the ONF (N_1, N_2) is free of torsion because there hold

$$T_{1,1}^2 = N_{1,u} \cdot N_2 = X_u \cdot N_2 = 0, \quad T_{1,2}^2 = N_{1,v} \cdot N_2 = X_v \cdot N_2 = 0.$$

2. A special example of such a spherical surface is the *flat Clifford torus,* given by the product (see e.g. do Carmo [36], Chap. 6)

$$X(u,v) = \frac{1}{\sqrt{2}}(\cos u, \sin u, \cos v, \sin v) \sim S^1 \times S^1.$$

We assign a moving 4-frame $\{X_u, X_v, N_1, N_2\}$ consisting of

$$X_u = \frac{1}{\sqrt{2}}(-\sin u, \cos u, 0, 0), \quad X_v = \frac{1}{\sqrt{2}}(0, 0, -\sin v, \cos v)$$

as well as

$$N_1 = \frac{1}{\sqrt{2}}(\cos u, \sin u, \cos v, \sin v), \quad N_2 = \frac{1}{\sqrt{2}}(-\cos u, -\sin u, \cos v, \sin v).$$

This special ONF $\mathfrak{N} = (N_1, N_2)$ is free of torsion. In another context, Pinl in [96] and [97] discusses examples of spherical surfaces already introduced by Killing, namely first

$$Y(u,v) = (\sin u \sin v, \cos u \cos v, \sin u \cos v, \cos u \sin v),$$

---

[1]In Sect. 2.1.2 we have already discussed surfaces with parallel mean curvature vector and a possible connection to the curvatures of their normal bundles.

a parametrization which was later considered again by Lawson in [85] in the more general form

$$(\cos \alpha u \cos v, \sin \alpha u \cos v, \cos u \sin v, \sin u \sin v)$$

and who proved that every ruled minimal surface in S^3 is of this form for some constant $\alpha > 0$, as well as

$$Z(u, v) = \frac{1}{2}(\cos u + \cos v, \sin u - \sin v, \sin u + \sin v, -\cos u + \cos v).$$

These three mappings X, Y and Z are spherical, and therefore they admit torsion-free orthonormal normal frames. Actually, they are flat since the first fundamental forms $\mathbf{g}$ are Euclidean, i.e.

$$\mathbf{g}(X) = \begin{pmatrix} 1 & 0 \\ 0 & 1 \end{pmatrix}, \quad \mathbf{g}(Y) = \begin{pmatrix} 2 & 0 \\ 0 & 2 \end{pmatrix}, \quad \mathbf{g}(Z) = \frac{1}{2}\begin{pmatrix} 1 & 0 \\ 0 & 1 \end{pmatrix}.$$

However, from the geometric point of view, Killing's examples have interesting projections into the three-dimensional subspaces.

3. Next, we want to consider so-called *parallel-type surfaces:* The immersion

$$R(u, v) = X(u, v) + f(u, v)N_1(u, v) + g(u, v)N_2(u, v)$$

is said to be *parallel to* X if the tangential planes of R and X are parallel at corresponding points. Whether surfaces in higher-dimensional spaces are parallel or not depends on the scalar curvature S of the normal bundle.

Proposition 3.2. *Let $f, g \neq 0$. If R is parallel to the immersion $X : \overline{B} \to \mathbb{R}^4$, i.e. if there hold*

$$R_{u^i} \cdot N_\sigma = 0 \quad \text{for all } i = 1, 2, \ \sigma = 1, 2,$$

at corresponding points, then $S \equiv 0$.

Proof. For the proof we use the Weingarten equations and compute the normal parts $R_u^\perp$ and $R_v^\perp$ of the tangential vectors R_u resp. R_v w.r.t. X, that is

$$R_u^\perp = f_u N_1 + g_u N_2 + f N_{1,u}^\perp + g N_{2,u}^\perp = \left(f_u - g T_{1,1}^2\right)N_1 + \left(g_u + f T_{1,1}^2\right)N_2$$

as well as

$$R_v^\perp = f_v N_1 + g_v N_2 + f N_{1,v}^\perp + g N_{2,v}^\perp = \left(f_v - g T_{1,2}^2\right)N_1 + \left(g_v + f T_{1,2}^2\right)N_2.$$

The condition that R is parallel to X then leads us to the first order system

$$f_u - gT_{1,1}^2 = 0, \quad f_v - gT_{1,2}^2 = 0,$$
$$g_u + fT_{1,1}^2 = 0, \quad g_v + fT_{1,2}^2 = 0.$$

Now differentiate the first two equations and make use of the others to get

$$0 = f_{uv} - g_v T_{1,1}^2 - g \partial_v T_{1,1}^2 = f_{uv} + fT_{1,1}^2 T_{1,2}^2 - g \partial_v T_{1,1}^2 ,$$
$$0 = f_{vu} - g_u T_{1,2}^2 - g \partial_u T_{1,2}^2 = f_{vu} + fT_{1,1}^2 T_{1,2}^2 - g \partial_u T_{1,2}^2 .$$

A comparison of the right hand sides shows

$$0 = -g \partial_v T_{1,1}^2 + g \partial_u T_{1,2}^2 = -g \cdot SW.$$

Similarly we find $0 = f \cdot SW$, and this proves the statement.   □

We will see that in case $S \equiv 0$ there is a torsion-free ONF $\mathfrak{N}$. Back to the first order system above, the proposition can be completed by stating $f, g \equiv$ const.

   Parallel-type surfaces are widely used in geometry and mathematical physics, see e.g. da Costa [30] for an application in quantum mechanics in curved spaces. In this context we would also like to refer the reader to the classical textbook Dirac [35]; for recent developments with applications in quantum string theory see e.g. Dorn et al. [38].

4. Finally we want to consider *evolute-type surfaces*. Again we consider the variation

$$T(u, v) = X(u, v) + f(u, v)N_1(u, v) + g(u, v)N_2(u, v).$$

The mapping T is then called an *evolute surface* to X if the tangential planes of T agree with the normal planes of X at corresponding points.

Proposition 3.3. *Let* $f, g \neq 0$. *If* T *from above is an evolute surface to the immersion* $X: \overline{B} \to \mathbb{R}^4$ *then* X *has flat normal bundle, i.e. it holds* $S \equiv 0$.

Proof. Using conformal parameters $(u, v) \in \overline{B}$ together with the Weingarten equations we compute the tangential parts of T_u and T_v, that is

$$T_u^{\mathsf{T}} = \left(1 - f\frac{L_{1,11}}{W} - g\frac{L_{2,11}}{W}\right) X_u - \left(f\frac{L_{1,12}}{W} + g\frac{L_{2,12}}{W}\right) X_v ,$$

$$T_v^{\mathsf{T}} = -\left(f\frac{L_{1,12}}{W} + g\frac{L_{2,12}}{W}\right) X_u + \left(1 - f\frac{L_{1,22}}{W} - g\frac{L_{2,22}}{W}\right) X_v .$$

Then the conditions

$$T_{u^i} \cdot X_{u^j} = 0 \quad \text{for } i, j = 1, 2$$

leads us to

$$W - fL_{1,11} - gL_{2,11} = 0, \quad W - fL_{1,22} - gL_{2,22} = 0,$$
$$fL_{1,12} + gL_{2,12} = 0.$$

Now use the Ricci integrability conditions to evaluate the scalar normal curvature

$$
\begin{aligned}
fgSW &= f(L_{1,11} - L_{1,22})gL_{2,12} - f(gL_{2,11} - gL_{2,22})L_{1,12} \\
&= -f^2(L_{1,11} - L_{1,22})L_{1,12} - f(W - fL_{1,11} - W + fL_{1,22})L_{1,12} \\
&= -f^2(L_{1,11} - L_{1,22})L_{1,12} + f^2(L_{1,11} - L_{1,22})L_{1,12} = 0.
\end{aligned}
$$

This proves the statement. $\qquad\square$

For this and further developments we refer to Cheshkova [25].

3.2 Normal Coulomb Frames

3.2.1 The Total Torsion

But what happens if $S \not\equiv 0$? Clearly, *there is no parallel frame,* and it is therefore desirable to construct ONF's with similar features. For this purpose we make the

Definition 3.1. The *total torsion* $\mathscr{T}[\mathfrak{N}]$ of an ONF $\mathfrak{N} = (N_1, N_2)$ is given by

$$\mathscr{T}[\mathfrak{N}] = \sum_{i,j=1}^{2} \sum_{\sigma,\vartheta=1}^{2} \iint_B g^{ij} T^{\vartheta}_{\sigma,i} T^{\vartheta}_{\sigma,j} W \, du \, dv.$$

Using conformal parameters with the properties

$$g^{11} = g^{22} = W^{-1} \quad \text{and} \quad g^{12} = 0,$$

and taking the skew-symmetry $T^{\vartheta}_{\sigma,i} = -T^{\sigma}_{\vartheta,i}$ of the torsion coefficients into account, reveals the *convex character of the functional of total torsion* for a fixed surface,

$$\mathscr{T}[\mathfrak{N}] = 2 \iint_B \left\{ (T^2_{1,1})^2 + (T^2_{1,2})^2 \right\} du \, dv.$$

We remark that $\mathscr{T}[\mathfrak{N}]$ *does not depend on the choice of the parametrization* (we only admit transformations which leave the unit normal vectors unaffected). But *it depends on the choice of the ONF* $\mathfrak{N}$, and it arises the question whether there are ONF's for which the functional of total torsion attains a smallest value.

3.2.2 Definition of Normal Coulomb Frames

Clearly, if the immersion X admits an ONF $\mathfrak{N}$ which is free of torsion, then it holds $\mathscr{T}[\mathfrak{N}] = 0$ for this special frame. On the other hand, we can make $\mathscr{T}[\mathfrak{N}]$ as large as we want by choosing φ sufficiently "bad."

Our goal is therefore to construct orthogonal unit normal frames which give $\mathscr{T}[\mathfrak{N}]$ a smallest possible value. Therefore our next

Definition 3.2. The ONF $\mathfrak{N} = (N_1, N_2)$ is called a *normal Coulomb frame* if it is critical for the functional $\mathscr{T}[\mathfrak{N}]$ of total torsion w.r.t. to $SO(2)$-valued variations of the form

$$\widetilde{N}_1 = \cos\varphi\, N_1 + \sin\varphi\, N_2\,, \quad \widetilde{N}_2 = -\sin\varphi\, N_1 + \cos\varphi\, N_2\,.$$

In the following we will compute and solve the Euler–Lagrange equation for normal Coulomb frames. We will furthermore establish various regularity properties for them. It particularly turns out that they are actually parallel, or free of torsion, in the special case of flat normal bundles.

3.2.3 The Euler–Lagrange Equation

We already computed in Sect. 3.1.2 the transformation between the torsion coefficients,

$$\widetilde{T}^2_{1,1} = T^2_{1,1} + \varphi_u\,, \quad \widetilde{T}^2_{1,2} = T^2_{1,2} + \varphi_v\,.$$

This gives us immediately the difference between the new and the old total torsion by means of a partial integration (see Sect. 3.2.1, use conformal parameters)

$$\mathscr{T}[\widetilde{\mathfrak{N}}] - \mathscr{T}[\mathfrak{N}] = 2\iint_B |\nabla\varphi|^2\, du\, dv + 4\iint_B (T^2_{1,1}\varphi_u + T^2_{1,2}\varphi_v)\, du\, dv$$

$$= 2\iint_B |\nabla\varphi|^2\, du\, dv + 4\int_{\partial B} (T^2_{1,1}, T^2_{1,2}) \cdot \nu\, \varphi\, ds$$

$$-4\iint_B \mathrm{div}\,(T^2_{1,1}, T^2_{1,2})\varphi\, du\, dv$$

with ν denoting the outer unit normal vector at the boundary ∂B.

Thus, we have proved the following criterion for a critical ONF $\mathfrak{N} = (N_1, N_2)$.

Proposition 3.4. *Let the ONF* $\mathfrak{N} = (N_1, N_2)$ *be critical for the functional* $\mathscr{T}[\mathfrak{N}]$ *of total torsion. Then, using conformal parameters* $(u, v) \in \overline{B}$, *the torsion coefficients* $T^{\vartheta}_{\sigma,i}$ *of this ONF satisfy the first order Neumann boundary value problem*

$$\text{div}\,(T^2_{1,1}, T^2_{1,2}) = 0 \quad \text{in } B, \quad (T^2_{1,1}, T^2_{1,2}) \cdot \nu = 0 \quad \text{on } \partial B$$

with the Euclidean divergence operator div.

The conservation law structure of this Euler–Lagrange equation explains the terminology *normal Coulomb frame* in analogy to Coulomb gauges from physics. We follow Helein [64] where he suggests to adopt this name from physics to differential geometry and harmonic analysis.

3.3   Constructing Normal Coulomb Frames, and Their Properties

3.3.1   *Construction via a Neumann Problem*

How can we construct a normal Coulomb frame $\mathfrak{N}$ from a given ONF $\widetilde{\mathfrak{N}}$? For a critical ONF $\mathfrak{N}$ we have to solve the boundary value problem

$$0 = \text{div}\,(T^2_{1,1}, T^2_{1,2}) = \text{div}\,(\widetilde{T}^2_{1,1} - \varphi_u, \widetilde{T}^2_{1,2} - \varphi_v) \quad \text{in } B,$$

$$0 = (T^2_{1,1}, T^2_{1,2}) \cdot \nu = (\widetilde{T}^2_{1,1} - \varphi_u, \widetilde{T}^2_{1,2} - \varphi_v) \cdot \nu \quad \text{on } \partial B$$

in virtue of the Euler–Lagrange equation from above. Therefore the

Proposition 3.5. *The given ONF $\widetilde{\mathfrak{N}}$ transforms into a normal Coulomb frame $\mathfrak{N}$ by means of the $SO(2)$-action if and only if*

$$\Delta\varphi = \text{div}\,(\widetilde{T}^2_{1,1}, \widetilde{T}^2_{1,2}) \quad \text{in } B,$$

$$\frac{\partial\varphi}{\partial\nu} = (\widetilde{T}^2_{1,1}, \widetilde{T}^2_{1,2}) \cdot \nu \quad \text{on } \partial B$$

holds true for the rotation angle $\varphi = \varphi(u, v)$.

What can we state about the solvability of this Neumann boundary value problem? We know that (see e.g. [29])

$$\Delta\varphi = f \quad \text{in } B, \quad \frac{\partial\varphi}{\partial\nu} = g \quad \text{on } \partial B$$

is solvable if and only if the following integrability condition holds

$$\iint_B f \, du\,dv = \int_{\partial B} g \, ds.$$

But this condition is obviously fulfilled in our situation:

$$\iint_B \operatorname{div}(\widetilde{T}^2_{1,1}, \widetilde{T}^2_{1,2})\, du\, dv = \int_{\partial B}(\widetilde{T}^2_{1,1}, \widetilde{T}^2_{1,2}) \cdot v\, ds.$$

Theorem 3.2. *Starting from a given ONF $\widetilde{\mathfrak{N}}$ it is always possible to construct a normal Coulomb frame $\mathfrak{N}$ for the immersion $X:\overline{B} \to \mathbb{R}^4$ which is critical for the functional $\mathcal{T}[\mathfrak{N}]$ of total torsion. In particular, if $X \in C^{k,\alpha}(\overline{B}, \mathbb{R}^4)$ with $k \geq 4$, then*

$$N_\sigma \in C^{k-1,\alpha}(\overline{B}, \mathbb{R}^4) \quad for \quad N_\sigma \in \mathfrak{N},\ \sigma = 1, 2.$$

For a general orientation on Neumann boundary value problems we refer the reader to the classical monograph of Courant and Hilbert [29]. Some detailed regularity analysis is also contained in our fourth chapter.

3.3.2 Minimality of Normal Coulomb Frames

Let $\mathfrak{N}$ be a normal Coulomb frame. Using conformal parameters we compute

$$\mathcal{T}[\widetilde{\mathfrak{N}}] = \mathcal{T}[\mathfrak{N}] + 2\iint_B |\nabla\varphi|^2\, du\, dv$$

$$+ 4\int_{\partial B}(T^2_{1,1}, T^2_{1,2}) \cdot v\, \varphi\, ds - 4\iint_B \operatorname{div}(T^2_{1,1}, T^2_{1,2})\varphi\, du\, dv$$

$$= \mathcal{T}[\mathfrak{N}] + 2\iint_B |\nabla\varphi|^2\, du\, dv \geq \mathcal{T}[\mathfrak{N}]$$

for a comparison ONF $\widetilde{\mathfrak{N}}$, taking into account that the boundary integral and the integral over the divergence vanish due to the Euler–Lagrange equation. From the parameter invariance of the functional of total torsion we then obtain

Theorem 3.3. *A normal Coulomb frame $\mathfrak{N}$ for the immersion $X:\overline{B} \to \mathbb{R}^4$ minimizes the functional of total torsion, i.e. it holds*

$$\mathcal{T}[\mathfrak{N}] \leq \mathcal{T}[\widetilde{\mathfrak{N}}]$$

for all ONF $\widetilde{\mathfrak{N}}$ resulting from a $SO(2)$-action applied to $\mathfrak{N}$. Equality occurs if and only if $\varphi \equiv$ const.

Note that a normal Coulomb frame remains $\mathcal{T}$-critical if it is rotated by a global constant rotation angle φ_0.

3.4   Estimating the Torsions of Normal Coulomb Frames

3.4.1   Reduction for Flat Normal Bundles

Let again $(u, v) \in \overline{B}$ be conformal parameters. We want to consider normal Coulomb frames in case of flat normal bundles $S \equiv 0$, that is

$$SW = \partial_v T_{1,1}^2 - \partial_u T_{1,2}^2 = \text{div}\,(-T_{1,2}^2, T_{1,1}^2) \equiv 0.$$

If $\mathfrak{N}$ is such a normal Coulomb frame then the Neumann boundary condition

$$(T_{1,1}^2, T_{1,2}^2) \cdot \nu = 0$$

from the Euler–Lagrange equation says *that the vector field* $(-T_{1,2}^2, T_{1,1}^2)$ *is parallel to the outer unit normal vector* ν *along* ∂B. Thus, a partial integration yields

$$\iint_B SW\,du\,dv = \int_{\partial B} (-T_{1,2}^2, T_{1,1}^2) \cdot \nu\,ds$$

$$= \pm \int_{\partial B} \sqrt{(T_{1,1}^2)^2 + (T_{1,2}^2)^2}\,ds.$$

In particular, $S \equiv 0$ implies

$$T_{\sigma,i}^{\vartheta} \equiv 0 \quad \text{on } \partial B \quad \text{for all } i = 1, 2 \text{ and } \sigma, \vartheta = 1, 2.$$

On the other hand, differentiate $0 = \partial_v T_{1,1}^2 - \partial_u T_{1,2}^2$ w.r.t. u and v and take the Euler–Lagrange equation $\partial_u T_{1,1}^2 + \partial_v T_{1,2}^2 = 0$ into account. Then we arrive at

$$\Delta T_{1,1}^2 = 0, \quad \Delta T_{1,2}^2 = 0$$

showing that $T_{1,1}^2$ and $T_{1,2}^2$ are harmonic functions. The maximum principle gives

$$T_{\sigma,i}^{\vartheta} \equiv 0 \quad \text{in } B \quad \text{for all } i = 1, 2 \text{ and } \sigma, \vartheta = 1, 2.$$

Theorem 3.4. *A normal Coulomb frame* $\mathfrak{N}$ *for an immersion* $X : \overline{B} \to \mathbb{R}^4$ *with flat normal bundle* $S \equiv 0$ *is always free of torsion. In other words, it is always parallel.*

To summarize our considerations so far: We have proved existence and regularity of ONF's which are critical for the functional of total torsion. If additionally the scalar curvature S of the normal bundle vanishes identically, such normal Coulomb frames are free of torsion.

3.4.2 Estimates via the Maximum Principle

Next we want to consider the case of *non-flat* normal bundles. From the Euler–Lagrange equation we know that the torsion vector $(T^2_{1,1}, T^2_{1,2})$ of a normal Coulomb frame is divergence-free. Thus *the differential* 1-*form*

$$\omega := -T^2_{1,2}\,du + T^2_{1,1}\,dv$$

is closed since for its exterior derivative we calculate

$$d\omega = -\partial_v T^2_{1,2}\,dv \wedge du + \partial_u T^2_{1,1}\,du \wedge dv = \mathrm{div}\,(T^2_{1,1}, T^2_{1,2})\,du \wedge dv = 0.$$

Poincaré's lemma therefore ensures the existence of a differentiable function τ satisfying (see e.g. Sauvigny [107], vol. 1, Chap. I, Sect. 7)

$$d\tau = \tau_u\,du + \tau_v\,dv = \omega.$$

This finally implies

$$\nabla\tau = (\tau_u, \tau_v) = (-T^2_{1,2}, T^2_{1,1}).$$

A second differentiation, taking account of $SW = \mathrm{div}\,(-T^2_{1,2}, T^2_{1,1})$, leads us to the following inhomogeneous boundary value problem

$$\Delta\tau = SW \quad \text{in } B, \quad \tau = 0 \quad \text{on } \partial B.$$

To justify the homogeneous boundary condition we note that

$$\nabla\tau \cdot (-v, u) = 0 \quad \text{on } \partial B$$

for normal Coulomb frames because $(T^2_{1,1}, T^2_{1,2})$ is perpendicular to the normal v. Therefore it holds $\tau = \text{const}$ along ∂B. But τ is only defined up to a constant of integration which can be chosen such that the homogeneous boundary condition is satisfied. Thus, Poisson's representation formula for the solution τ reads

$$\tau(w) = \iint\limits_B \Phi(\zeta; w)S(\zeta)W(\zeta)\,d\xi d\eta, \quad \zeta = (\xi, \eta) \in \overline{B},$$

with the non-positive Green kernel of the Laplace operator Δ (see e.g. Sauvigny [107], vol. 2, Chap. VIII, Sect. 1). We want to give an integral estimate for this kernel to establish an estimate for the integral function τ : Namely, note that

$$\psi(w) = \frac{|w|^2 - 1}{4} \quad \text{solves} \quad \Delta\psi = 1 \text{ in } B \quad \text{and} \quad \psi = 0 \text{ on } \partial B.$$

Therefore we conclude

$$\iint\limits_{B} |\Phi(\zeta; w)| \cdot 1 \, d\xi d\eta = \frac{1 - |w|^2}{4} \le \frac{1}{4}.$$

Lemma 3.2. *Let $(u, v) \in \overline{B}$ be conformal parameters. The integral function τ of the above Poisson problem with the scalar curvature S of the normal bundle satisfies*

$$|\tau(w)| \le \frac{1}{4} \|SW\|_{C^0(\overline{B})} \quad in \ B, \quad \tau(w) = 0 \quad on \ \partial B.$$

Well-known potential theoretic estimates for the Laplacian (see e.g. [107], Chap. IX, Sect. 4, Satz 1) now ensure the existence of two constants $c_1 \in (0, \infty)$ and $c_2(\alpha) \in (0, \infty)$ such that

$$\|\tau\|_{C^1(\overline{B})} \le c_1 \|SW\|_{C^0(\overline{B})}$$

holds true, or also an estimate for the higher derivative

$$\|\tau\|_{C^{2+\alpha}(\overline{B})} \le c_2(\alpha) \|SW\|_{C^\alpha(\overline{B})}$$

for all $\alpha \in (0, 1)$.[2] With the foregoing C^1-bound we can formulate the main result of the present chapter.

Theorem 3.5. *Let the conformally parametrized immersion $X: \overline{B} \rightarrow \mathbb{R}^4$ with normal bundle of scalar curvature S be given. Then there exists a normal Coulomb frame $\mathfrak{N}$ minimizing the functional of total torsion, with torsion coefficients satisfying*

$$\|T^\vartheta_{\sigma,i}\|_{C^0(\overline{B})} \le C_1 \|SW\|_{C^0(\overline{B})} \quad and \quad \|T^\vartheta_{\sigma,i}\|_{C^{1+\alpha}(\overline{B})} \le C_2(\alpha) \|SW\|_{C^\alpha(\overline{B})}$$

for all $\alpha \in (0, 1)$ with real constants $C_1 \in (0, \infty)$ and $C_2(\alpha) \in (0, \infty)$.

In particular, *for flat normal bundles with $S \equiv 0$ we recover that normal Coulomb frames are free of torsion.*

3.4.3 Estimates via a Cauchy–Riemann Boundary Value Problem

Consider again the Euler–Lagrange equation for a normal Coulomb frame together with the formula for the scalar curvature of the normal bundle, that is

$$\partial_u T^2_{1,1} + \partial_v T^2_{1,2} = 0, \quad \partial_v T^2_{1,1} - \partial_u T^2_{1,2} = SW.$$

---

[2]The constants c_1, c_2 and C_1, C_2 below depend also on the domain $\overline{B}$.

We want to discuss a second method to control the torsion coefficients of a normal Coulomb frame which is strongly adapted to the case of two codimensions: Namely, we will prove that a complex-valued torsion Ψ from the next lemma satisfies a first-order differential equation which can be solved using methods from Vekua [117]. Theorem 3.5 will be replaced by an estimate of certain L^p-norms of SW.

Lemma 3.3. *Let $\mathfrak{N}$ be a normal Coulomb frame. Then the complex-valued torsion*

$$\Psi := T^2_{1,1} - i T^2_{1,2} \in \mathbb{C}$$

solves the inhomogeneous Cauchy–Riemann equation

$$\Psi_{\overline{w}} = \frac{i}{2} SW \quad \text{in } B,$$

$$\text{Re}\big[w\Psi(w)\big] = 0 \quad \text{for } w \in \partial B$$

using conformal parameters $w = (u, v) \in \overline{B}$ with the setting $\Phi_{\overline{w}} := \frac{1}{2}(\Phi_u + i\Phi_v)$.

Proof. We compute

$$2\Psi_{\overline{w}} = \Psi_u + i\Psi_v = (\partial_u T^2_{1,1} + \partial_v T^2_{1,2}) + i(\partial_v T^2_{1,1} - \partial_u T^2_{1,2}) = 0 + iSW \in \mathbb{C}$$

as well as

$$\text{Re}\big[w\Psi(w)\big] = \text{Re}\big[(u + iv)(T^2_{1,1} - iT^2_{1,2})\big] = uT^2_{1,1} + vT^2_{1,2}$$
$$= (-T^2_{1,2}, T^2_{1,1}) \cdot (-v, u) = \nabla\tau \cdot (-v, u) = 0$$

with the integral function τ from the previous paragraph.   □

We say that Ψ *solves a linear Riemann–Hilbert problem.* There is a huge complex analysis machinery to treat such a mathematical problem. The reader is referred e.g. to Bers [11], Begehr [7], Courant and Hilbert [29], Sauvigny [107], Vekua [117], Wendland [121].

Let us now derive an integral representation for the complex-valued torsion Ψ.

Lemma 3.4. *The above Riemann–Hilbert problem for the complex-valued torsion Ψ of a normal Coulomb frame possesses at most one solution $\Psi \in C^1(B, \mathbb{C}) \cap C^0(\overline{B}, \mathbb{C})$.*

Proof. Let Ψ_1, Ψ_2 be two such solutions. Then we set

$$\Phi(w) := w[\Psi_1(w) - \Psi_2(w)]$$

and compute immediately

$$\Phi_{\overline{w}} = 0 \quad \text{in } B, \qquad \text{Re}\,\Phi = 0 \quad \text{on } \partial B.$$

Thus the real part of the holomorphic function Φ vanishes on ∂B, and the Cauchy–Riemann equations yield $\Phi \equiv ic$ in B with some constant $c \in \mathbb{R}$. The continuity of Ψ_1 and Ψ_2 implies $c = 0$ since $\Phi(0,0) = (0,0)$. $\qquad\qquad\qquad\qquad\qquad\square$

Our Riemann–Hilbert problem can now be solved by means of so-called *generalized analytic functions* for which we want to present some basic facts (see the references above for more comprehensive presentations).

For arbitrary $f \in C^1(\overline{B}, \mathbb{C})$ we define *Cauchy's integral operator* by

$$\mathscr{T}_B[f](w) := -\frac{1}{\pi} \iint_B \frac{f(\zeta)}{\zeta - w}\, d\xi d\eta, \quad w \in \mathbb{C}.$$

Lemma 3.5. *There hold $\mathscr{T}_B[f] \in C^1(\mathbb{C} \setminus \partial B, \mathbb{C}) \cap C^0(\mathbb{C}, \mathbb{C})$ as well as*

$$\frac{\partial}{\partial \overline{w}}\, \mathscr{T}_B[f](w) = \begin{cases} f(w), & w \in B \\ 0, & w \in \mathbb{C} \setminus \overline{B} \end{cases}.$$

Proof. For a detailed proof see e.g. Vekua [117], Chap. I, Sect. 5. We only verify the complex derivative: Let $\{G_k\}_{k=1,2,...}$ be a sequence of open, simply connected and smoothly bounded domains contracting to some point $z_0 \in B$ for $k \to \infty$. Let $|G_k|$ denote its area. With the characteristic function χ we compute (see [107], Chap. IV, Sect. 5)

$$\frac{1}{2i|G_k|} \int_{\partial G_k} \mathscr{T}_B[f](w)\, dw = \frac{1}{2i|G_k|} \int_{\partial G_k} \left(-\frac{1}{\pi} \iint_B \frac{f(\zeta)}{\zeta - w}\, d\xi d\eta \right) dw$$

$$= \frac{1}{2\pi i|G_k|} \iint_B \left(f(\zeta) \int_{\partial G_k} \frac{1}{w - \zeta}\, dw \right) d\xi d\eta$$

$$= \frac{1}{2\pi i|G_k|} \iint_B f(\zeta) \cdot 2\pi i \chi_{G_k}(\zeta)\, d\xi d\eta$$

$$= \frac{1}{|G_k|} \cdot \iint_{G_k} f(\zeta)\, d\xi d\eta.$$

Here we have used Cauchy's integral formula

$$\int_{\partial G_k} \frac{g(w)}{w - \zeta}\, dw = 2\pi i g(\zeta)$$

for a holomorphic function g with $\zeta \in G_k$. Applying the integration by parts rule in complex form

$$\iint\limits_{G_k} \frac{d}{d\overline{w}} f(w) \, d\xi d\eta = \frac{1}{2i} \int\limits_{\partial G_k} f(z) \, dz$$

we get in the limit

$$\frac{d}{d\overline{w}} \mathscr{T}_B[f](w) = \lim_{k \to \infty} \frac{1}{2i|G_k|} \int\limits_{\partial G_k} \mathscr{T}_B[f](w) \, dw = f(z_0)$$

for all $z_0 \in B$. The statement follows. $\qquad\qquad\square$

Next we set

$$\mathscr{P}_B[f](w) := -\frac{1}{\pi} \iint\limits_B \left\{ \frac{f(\zeta)}{\zeta - w} + \frac{\overline{\zeta} \, \overline{f(\zeta)}}{1 - w\overline{\zeta}} \right\} d\xi \, d\eta = \mathscr{T}_B[f](w) - \frac{1}{w} \overline{\mathscr{T}_B[wf]\left(\frac{1}{\overline{w}}\right)}.$$

Then Satz 1.24 in Vekua [117] states the following

Lemma 3.6. *With the definitions above we have the uniform estimate*

$$\left| \mathscr{P}_B[f](w) \right| \le C(p) \|f\|_{L^p(B)}, \quad w \in B,$$

where $p \in (2, +\infty]$, and $C(p)$ is a positive constant depending on p.

Using this result (which remains unproved here) we obtain our second torsion estimate, now in terms of L^p-norms, the main result of this paragraph.

Theorem 3.6. *Let the conformally parametrized immersion $X : \overline{B} \to \mathbb{R}^4$ be given. Then the complex-valued torsion Ψ of a normal Coulomb frame $\mathfrak{N}$ satisfies*

$$|\Psi(w)| \le c(p) \|SW\|_{L^p(B)} \quad \text{for all } w \in B$$

with a positive constant $c(p)$, where $p \in (2, +\infty]$.

For a flat normal bundle with $S \equiv 0$ we thus verify again our results from Sect. 3.4.1.

Proof. Let us abbreviatory write

$$f := \frac{i}{2} SW \in C^1(\overline{B}, \mathbb{C})$$

to apply the previous results. We claim that the complex-valued torsion Ψ possesses the integral representation

$$\Psi(w) = \mathscr{P}_B[f](w) = -\frac{1}{\pi} \iint\limits_B \left\{ \frac{f(\zeta)}{\zeta - w} + \frac{\overline{\zeta}\,\overline{f(\zeta)}}{1 - w\overline{\zeta}} \right\} d\xi\, d\eta, \quad w \in B.$$

Then the stated estimate follows at once from the above lemma.

1. First we claim

$$w\mathscr{P}_B[f](w) = \frac{1}{\pi} \iint\limits_B f(\zeta)\, d\xi\, d\eta + \mathscr{T}_B[wf](w) - \overline{\mathscr{T}_B[wf]\left(\frac{1}{\overline{w}}\right)}.$$

Let us check this identity:

$$\frac{1}{\pi} \iint\limits_B f(\zeta)\, d\xi d\eta + \mathscr{T}_B[wf](w) - \overline{\mathscr{T}_B[wf](\overline{w}^{-1})}$$

$$= \frac{1}{\pi} \iint\limits_B f(\zeta)\, d\xi d\eta - \frac{1}{\pi} \iint\limits_B \frac{\zeta f(\zeta)}{\zeta - w}\, d\xi d\eta - \overline{\mathscr{T}_B[wf](\overline{w}^{-1})}$$

$$= \cdots\cdots\cdots\cdots\cdots\cdots\cdots\cdots\cdots\cdots\cdots\cdots\cdots\cdots\cdots\cdots$$

$$\cdots\cdots = -\frac{w}{\pi} \iint\limits_B \frac{f(\zeta)}{\zeta - w}\, d\xi d\eta + \frac{1}{\pi} \iint\limits_B \frac{\overline{\zeta}\,\overline{f(\zeta)}}{\overline{\zeta} - \frac{1}{w}}\, d\xi d\eta$$

$$= -\frac{w}{\pi} \iint\limits_B \frac{f(\zeta)}{\zeta - w}\, d\xi d\eta + \frac{w}{\pi} \iint\limits_B \frac{\overline{\zeta}\,\overline{f(\zeta)}}{\overline{\zeta} w - 1}\, d\xi d\eta$$

$$= -\frac{w}{\pi} \iint\limits_B \left(\frac{f(\zeta)}{\zeta - w} + \frac{\overline{\zeta}\,\overline{f(\zeta)}}{1 - \overline{\zeta} w} \right) d\xi d\eta.$$

2. Next, taking $f = \frac{i}{2} SW$ into account, we infer

$$\mathrm{Re}\left\{ w\mathscr{P}_B[f](w) \right\} = 0, \quad w \in \partial B,$$

what follows from

$$\mathscr{T}_B[\tfrac{1}{2} iwSW](w) - \overline{\mathscr{T}_B[\tfrac{1}{2} iwSW](\overline{w}^{-1})}$$

$$= -\frac{1}{\pi} \iint\limits_B \frac{i}{2} \frac{\zeta SW}{\zeta - w}\, d\xi d\eta + \frac{1}{\pi} \iint\limits_B \frac{i}{2} \frac{\zeta SW}{\zeta - \overline{w}^{-1}}$$

$$= -\frac{i}{2\pi} \iint\limits_B \left(\frac{\zeta}{\zeta - w} + \frac{\overline{\zeta}}{\overline{\zeta} - \frac{1}{w}} \right) SW\, d\xi d\eta.$$

The entry in the brackets is a real number since it holds

$$\frac{1}{w} = \frac{\overline{w}}{|w|^2} = \overline{w} \quad \text{on } \partial B.$$

Investing

$$\frac{\partial}{\partial \overline{w}} \, \mathscr{P}_B[f](w) = f(w),$$

which follows from Lemma 3.5 and our representation of $\mathscr{P}_B[f](w)$, we conclude that $\mathscr{P}_B[f](w)$ solves the Riemann–Hilbert problem for Ψ. The above uniqueness result for the Riemann–Hilbert problem proves the stated representation. $\square$

3.4.4   A Gradient Estimate for Torsion Vectors of Constant Length

The question remains how we can prove *pointwise estimates* for the torsion coefficients $T_{\sigma,i}^{\vartheta}(w)$ of normal Coulomb frames in terms of the scalar normal curvature $S(w)$. At least we want to present the following special result concerning the gradients of the $T_{\sigma,i}^{\vartheta}$ for torsion vectors of constant length.

Theorem 3.7. *Let the conformally parametrized immersion $X \colon \overline{B} \to \mathbb{R}^4$ together with a normal Coulomb frame $\mathfrak{N}$ be given. Assume that there hold*

$$T_{1,1}^2 \neq 0, \ T_{1,2}^2 \neq 0 \quad \text{and} \quad |T_{1,1}^2|^2 + |T_{1,2}^2|^2 = \text{const} \quad \text{in } \overline{B}.$$

Then there hold the pointwise estimates

$$|\partial_{u^k} T_{\sigma,i}^{\vartheta}(w)| \leq |S(w)| W(w) \quad \text{for all } w \in B$$

and for all $i, k = 1, 2$ and $\sigma, \vartheta = 1, 2$.

Proof. For abbreviation we set

$$f := T_{1,1}^2, \quad g := T_{1,2}^2.$$

Then the Euler–Lagrange equation together with the definition of the scalar normal curvature S read

$$f_u + g_v = 0, \quad f_v - g_u = SW \quad \text{in } B.$$

Moreover, differentiating $f^2 + g^2 = \text{const}$ w.r.t u and v yields

$$f_u g_v - f_v g_u = 0 \quad \text{in } B$$

taking the assumption $f, g \neq 0$ into account. We arrive at

$$\left. \begin{array}{ll} f_u^2 + f_u g_v = 0, & f_v^2 - f_v g_u = SW f_v \\ f_u g_v + g_v^2 = 0, & f_v g_u - g_u^2 = SW g_u \end{array} \right\} \quad \text{implying} \quad \begin{array}{l} |\nabla f|^2 = SW f_v \\ |\nabla g|^2 = -SW g_u \end{array}.$$

The Cauchy–Schwarz inequality thus gives

$$|\nabla f|^2 \leq \frac{1}{2} (SW)^2 + \frac{1}{2} f_v^2 \leq \frac{1}{2} (SW)^2 + \frac{1}{2} |\nabla f|^2 ,$$

and an analogous estimate holds true for $|\nabla g|^2$. We finally infer

$$|\nabla f|^2 \leq (SW)^2 , \quad |\nabla g|^2 \leq (SW)^2 \quad \text{in } B,$$

and the statement follows. $\qquad\qquad\qquad\qquad\qquad\qquad\qquad\qquad\qquad\qquad$ □

3.4.5 Estimates for the Total Torsion

The previous results allow us to establish immediately lower and upper bounds for the total torsion $\mathscr{T}[\mathfrak{N}]$ of normal Coulomb frames $\mathfrak{N}$. In particular, Theorems 3.5 and 3.6 show

Theorem 3.8. *Let the conformally parametrized immersion $X : \overline{B} \to \mathbb{R}^4$ with a normal Coulomb frame $\mathfrak{N}$ be given.*
Then there hold

$$\mathscr{T}[\mathfrak{N}] \leq 4 C_1^2 \pi \| SW \|_{C^0(\overline{B})}^2$$

with the real constant $C_1 \in (0, \infty)$ from Theorem 3.5, as well as

$$\mathscr{T}[\mathfrak{N}] \leq 2 c(p)^2 \pi \| SW \|_{L^p(B)}^2$$

for all $p \in (2, +\infty]$ with the real constant $c = c(p)$ from Theorem 3.6.

The question remains open whether there even exist an integral estimate of the form

$$\mathscr{T}[\mathfrak{N}] \leq \gamma_1 \iint\limits_B |S|^q W \, du \, dv$$

with some sufficiently chosen constant $\gamma_1 \in (0, \infty)$ and $q \geq 1$. Notice, as already mentioned briefly in part, that the difficulties are hidden behind pointwise or higher-order L^p-estimates of S and W.

The following lower bound for the total torsion of normal Coulomb frames $\mathfrak{N}$ is a special case of a general estimate which we will prove later when we consider the general case of higher codimension. We will therefore only state the result without giving a proof.

Theorem 3.9. *Let the conformally parametrized immersion $X: \overline{B} \to \mathbb{R}^4$ with a normal Coulomb frame $\mathfrak{N}$ be given. Assume $S \not\equiv 0$ and $\|\nabla S\|^2_{L^2(B)} > 0$ for the scalar curvature S of its normal bundle. Then it holds*

$$\mathscr{T}[\mathfrak{N}] \geq \left(\frac{\|S\|^2_{L^2(B)}}{(1-\rho)^2 \mathscr{S}_2[X; B_\rho]} + \frac{2\|\nabla S\|^2_{L^2(B)}}{\mathscr{S}_2[X; B_\rho]} \right)^{-1} \mathscr{S}_2[X; B_\rho]$$

with the functional

$$\mathscr{S}_2[X; B_\rho] := \iint\limits_{B_\rho} |S|^2 W \, du \, dv,$$

and where the radius $\rho \in (0, 1)$ is chosen such that $\mathscr{S}_2[X; B_\rho] > 0$.

3.5　An Example: Holomorphic Graphs

We consider again graphs

$$X(w) = \big(w, \Phi(w)\big), \quad w = u + iv \in B,$$

with a holomorphic function $\Phi = \varphi + i\psi$. We want to show that its Euler unit normal vectors represent a normal Coulomb frame if it holds $|\Phi| = \text{const}$ along the boundary curve ∂B. This is true e.g. for the choice $\Phi(w) = w^n$.
First, there hold

$$g_{11} = 1 + |\nabla \varphi|^2 = g_{22}, \quad g_{12} = 0$$

due to the Cauchy–Riemann equations $\varphi_u = \psi_v, \varphi_v = -\psi_u$, therefore also

$$\Delta \varphi = \Delta \psi = 0,$$

i.e. the graph $X = X(u, v)$ is conformally parametrized and represents a minimal surface in $\mathbb{R}^4$ satisfying

$$\Delta X(u, v) = 0 \quad \text{in } B.$$

Its area element $W > 0$ is given by

$$W = 1 + |\nabla \varphi|^2 = 1 + |\nabla \psi|^2,$$

and its Euler unit normal vectors are

$$N_1 = \frac{1}{\sqrt{W}} \, (-\varphi_u, -\varphi_v, 1, 0), \quad N_2 = \frac{1}{\sqrt{W}} \, (-\psi_u, -\psi_v, 0, 1).$$

For the associated torsion coefficients we have

$$T_{1,1}^2 = \frac{1}{W} \, (-\varphi_{uu}\varphi_v + \varphi_{uv}\varphi_u) = \frac{1}{2W} \frac{\partial}{\partial v} |\nabla\varphi|^2, \quad T_{1,2}^2 = -\frac{1}{2W} \frac{\partial}{\partial u} |\nabla\varphi|^2.$$

Consequently, due to the special form of W we infer

$$\text{div}\,(T_{1,1}^2, T_{1,2}^2) = \frac{\partial}{\partial u} \left(\frac{1}{2W} \right) \frac{\partial}{\partial v} |\nabla\varphi|^2 - \frac{\partial}{\partial v} \left(\frac{1}{2W} \right) \frac{\partial}{\partial u} |\nabla\varphi|^2$$

$$+ \frac{1}{2W} \frac{\partial^2}{\partial v \partial u} |\nabla\varphi|^2 - \frac{1}{2W} \frac{\partial^2}{\partial u \partial v} |\nabla\varphi|^2$$

$$= \frac{1}{2W^2} \frac{\partial}{\partial u} |\nabla\varphi|^2 \frac{\partial}{\partial v} |\nabla\varphi|^2 - \frac{1}{2W^2} \frac{\partial}{\partial v} |\nabla\varphi|^2 \frac{\partial}{\partial u} |\nabla\varphi|^2 = 0.$$

Thus the Euler–Lagrange equation for a normal Coulomb frame is satisfied.

In order to check the boundary condition from the Euler–Lagrange equation for the total torsion we introduce polar coordinates $u = r \cos\omega$, $v = r \sin\omega$. Note that

$$\frac{1}{r} \frac{\partial}{\partial \omega} = u \frac{\partial}{\partial v} - v \frac{\partial}{\partial u}.$$

In our case we obtain

$$(T_{1,1}^2, T_{1,2}^2) \cdot \nu = \frac{1}{2W} \left(u \frac{\partial}{\partial v} - v \frac{\partial}{\partial u} \right) |\nabla\varphi|^2 = \frac{1}{2W} \frac{\partial}{\partial \omega} |\Phi_w|^2 \quad \text{on } \partial B$$

with the complex derivative $\Phi_w = \frac{1}{2} (\Phi_u - i \Phi_v)$.
We infer that the boundary condition from the Euler–Lagrange equation is satisfied if and only if $\frac{\partial}{\partial \omega} |\Phi_w|$ vanishes identically at the boundary curve ∂B.

Thus we have proved

Proposition 3.6. *Let the conformally parametrized minimal graph $(w, \Phi(w))$ with a holomorphic function $\Phi = \varphi + i\psi$ be given. Then its unit Euler normal vectors N_1 and N_2 form a normal Coulomb frame $\mathfrak{N}$ if Φ_w does not depend on the angle ω along the boundary curve ∂B.*

As mentioned, this result applies to our well-known examples

$$X(w) = (w, w^n) \quad \text{with } n \in \mathbb{N}.$$

Chapter 4
Normal Coulomb Frames in $\mathbb{R}^{n+2}$

Abstract Now we consider two-dimensional surfaces immersed in Euclidean spaces $\mathbb{R}^{n+2}$ of arbitrary dimension. The construction of normal Coulomb frames turns out to be more intricate and requires a profound analysis of nonlinear elliptic systems in two variables.

The Euler–Lagrange equations of the functional of total torsion are identified as non-linear elliptic systems with quadratic growth in the gradient, and, more exactly, the nonlinearity in the gradient is of so-called *curl-type*, while the Euler–Lagrange equations appear in a *div-curl-form*.

We discuss the interplay between curvatures of the normal bundles and torsion properties of normal Coulomb frames. It turns out that such frames are free of torsion if and only if the normal bundle is flat.

Existence of normal Coulomb frames is then established by solving a variational problem in a weak sense using ideas of F. Helein (Harmonic Maps, Conservation Laws and Moving Frames, Cambridge University Press, Cambridge, 2002). This, of course, ensures minimality, but we are also interested in classical regularity of our frames. For this purpose we employ deep results of the theory of nonlinear elliptic systems of div-curl-type and benefit from the work of many authors: E. Heinz, S. Hildebrandt, F. Helein, F. Müller, S. Müller, T. Rivière, F. Sauvigny, A. Schikorra, E.M. Stein, F. Tomi, H.C. Wente, and many others.

4.1 Problem Formulation

In this fourth chapter we want to generalize the previous considerations to the case of higher codimensions $n \geq 2$. We start with computing the Euler–Lagrange equations for the functional of total torsion

$$\mathscr{T}[\mathfrak{N}] = \sum_{i,j=1}^{2} \sum_{\sigma,\vartheta=1}^{n} g^{ij} T_{\sigma,i}^{\vartheta} T_{\sigma,j}^{\vartheta} W \, du \, dv$$

S. Fröhlich, *Coulomb Frames in the Normal Bundle of Surfaces in Euclidean Spaces*, Lecture Notes in Mathematics 2053, DOI 10.1007/978-3-642-29846-2_4, © Springer-Verlag Berlin Heidelberg 2012

for an ONF $\mathfrak{N} = (N_1, \ldots, N_n)$. It turns out that these equations form a non-linear elliptic system with quadratic growth in the gradient.

We will prove existence and regularity of parallel normal frames in case of vanishing curvature of the normal bundles as well as of critical points of $\mathscr{T}[\mathfrak{N}]$ in the general situation of non-flat normal bundles. Our main intention is to discuss analytical and geometrical properties of such normal Coulomb frames.

4.2   The Euler–Lagrange Equations

4.2.1   Definition of Normal Coulomb Frames

In this section we derive the Euler–Lagrange equations for $\mathscr{T}$-critical ONF's $\mathfrak{N}$. For this purpose we first mention that due to do Carmo [36], Chap. 3 we can construct a family $\mathbf{R}(w, \varepsilon) \in SO(n)$ of rotations for a given skew-symmetric matrix $\mathbf{A}(w) \in C^\infty(B, \mathrm{so}(n))$ by means of the geodesic flow in the manifold $SO(n)$.

In terms of such rotations we consider variations $\widetilde{\mathfrak{N}}$ of a given ONF $\mathfrak{N}$ by means of

$$\widetilde{N}_\sigma(w, \varepsilon) := \sum_{\vartheta=1}^{n} R_\sigma^\vartheta(w, \varepsilon) N_\vartheta(w), \quad \sigma = 1, \ldots, n.$$

To be precise, the family of rotations is given by

$$\mathbf{R}(w, \varepsilon) = \left(R_\sigma^\vartheta(w, \varepsilon) \right)_{\sigma, \vartheta = 1, \ldots, n} \in C^\infty(B \times (-\varepsilon_0, +\varepsilon_0), SO(n)),$$

with sufficiently small $\varepsilon_0 > 0$ such that

$$\mathbf{R}(w, 0) = \mathbb{E}^n, \quad \frac{\partial}{\partial \varepsilon} \mathbf{R}(w, 0) = \mathbf{A}(w) \in C^\infty(B, \mathrm{so}(n))$$

hold true with the n-dimensional unit matrix $\mathbb{E}^n$. Such a matrix $\mathbf{A}(w)$ is the essential ingredient for defining the first variation of the functional of total torsion.

Definition 4.1.   An ONF $\mathfrak{N}$ is called *critical for the total torsion* or shortly a *normal Coulomb frame* if and only if the first variation

$$\delta_\mathbf{A} \mathscr{T}[\mathfrak{N}] := \lim_{\varepsilon \to 0} \frac{1}{\varepsilon} \left\{ \mathscr{T}[\widetilde{\mathfrak{N}}] - \mathscr{T}[\mathfrak{N}] \right\}$$

vanishes w.r.t. all skew-symmetric perturbations $\mathbf{A}(w) \in C^\infty(B, \mathrm{so}(n))$.

4.2.2 The First Variation

We want to compute the Euler–Lagrange equations for normal Coulomb frames in the sense of the foregoing definition.

Proposition 4.1. *The ONF $\mathfrak{N}$ is a normal Coulomb frame for the conformally parametrized immersion $X \colon B \to \mathbb{R}^{n+2}$ if and only if its torsion coefficients solve the following system of Neumann boundary value problems*

$$\operatorname{div}\left(T_{\sigma,1}^{\vartheta}, T_{\sigma,2}^{\vartheta}\right) = 0 \quad \text{in } B, \quad \left(T_{\sigma,1}^{\vartheta}, T_{\sigma,2}^{\vartheta}\right) \cdot v = 0 \quad \text{on } \partial B$$

for all $\sigma, \vartheta = 1, \ldots, n$, where v is the outer unit normal vector along ∂B.

As already mentioned in the previous chapter, the conservation law structure of these Euler–Lagrange equations motivated us to use the terminology "Coulomb frame," and with this we actually follow a suggestion of Helein [64].

Proof. Use conformal parameters and consider a one-parameter family of rotations

$$\mathbf{R}(w, \varepsilon) = \left(R_{\sigma}^{\vartheta}(w, \varepsilon)\right)_{\sigma, \vartheta = 1, \ldots, n} \in SO(n)$$

as above. Expanding around $\varepsilon = 0$ yields

$$\mathbf{R}(w, \varepsilon) = \mathbb{E}^{n} + \varepsilon \mathbf{A}(w) + o(\varepsilon).$$

Applying $\mathbf{R}(w, \varepsilon)$ to the ONF $\mathfrak{N}$ gives the new ONF $\widetilde{\mathfrak{N}}$ in the form

$$\widetilde{N}_{\sigma} = \sum_{\vartheta=1}^{n} R_{\sigma}^{\vartheta} N_{\vartheta} = \sum_{\vartheta=1}^{n} \left\{\delta_{\sigma}^{\vartheta} + \varepsilon A_{\sigma}^{\vartheta} + o(\varepsilon)\right\} N_{\vartheta} = N_{\sigma} + \varepsilon \sum_{\vartheta=1}^{n} A_{\sigma}^{\vartheta} N_{\vartheta} + o(\varepsilon).$$

Now we compute

$$\widetilde{N}_{\sigma,u^{\ell}} = N_{\sigma,u^{\ell}} + \varepsilon \sum_{\vartheta=1}^{n} \left(A_{\sigma,u^{\ell}}^{\vartheta} N_{\vartheta} + A_{\sigma}^{\vartheta} N_{\vartheta,u^{\ell}}\right) + o(\varepsilon)$$

for the derivatives of these new unit normal vectors. Consequently, the new torsion coefficients can be expanded to

$$\widetilde{T}_{\sigma,\ell}^{\omega} = \widetilde{N}_{\sigma,u^{\ell}} \cdot \widetilde{N}_{\omega}$$

$$= N_{\sigma,u^{\ell}} \cdot N_{\omega} + \varepsilon \sum_{\vartheta=1}^{n} \left(A_{\sigma,u^{\ell}}^{\vartheta} N_{\vartheta} + A_{\sigma}^{\vartheta} N_{\vartheta,u^{\ell}}\right) \cdot N_{\omega} + \varepsilon N_{\sigma,u^{\ell}} \cdot \sum_{\vartheta=1}^{n} A_{\omega}^{\vartheta} N_{\vartheta} + o(\varepsilon)$$

$$= T_{\sigma,\ell}^{\omega} + \varepsilon A_{\sigma,u^{\ell}}^{\omega} + \varepsilon \sum_{\vartheta=1}^{n} \left\{A_{\sigma}^{\vartheta} T_{\vartheta,\ell}^{\omega} + A_{\omega}^{\vartheta} T_{\sigma,\ell}^{\vartheta}\right\} + o(\varepsilon),$$

and for their squares we infer

$$(\widetilde{T}^\omega_{\sigma,\ell})^2 = (T^\omega_{\sigma,\ell})^2 + 2\varepsilon \left\{ A^\omega_{\sigma,u\ell} T^\omega_{\sigma,\ell} + \sum_{\vartheta=1}^n \left(A^\vartheta_\sigma T^\omega_{\vartheta,\ell} T^\omega_{\sigma,\ell} + A^\vartheta_\omega T^\vartheta_{\sigma,\ell} T^\omega_{\sigma,\ell} \right) \right\} + o(\varepsilon).$$

Before we insert this identity into the functional $\mathscr{T}[\mathfrak{N}]$ we observe

$$\sum_{\sigma,\omega,\vartheta=1}^n \left\{ A^\vartheta_\sigma T^\omega_{\vartheta,\ell} T^\omega_{\sigma,\ell} + A^\vartheta_\omega T^\vartheta_{\sigma,\ell} T^\omega_{\sigma,\ell} \right\} = \sum_{\sigma,\omega,\vartheta=1}^n \left\{ A^\vartheta_\sigma T^\omega_{\vartheta,\ell} T^\omega_{\sigma,\ell} + A^\vartheta_\sigma T^\vartheta_{\omega,\ell} T^\sigma_{\omega,\ell} \right\}$$

$$= 2 \sum_{\sigma,\omega,\vartheta=1}^n A^\vartheta_\sigma T^\omega_{\vartheta,\ell} T^\omega_{\sigma,\ell} = 0$$

taking the skew-symmetry of the matrix $\mathbf{A}(w)$ into account. Thus the difference between $\mathscr{T}[\widetilde{\mathfrak{N}}]$ and $\mathscr{T}[\mathfrak{N}]$ computes to (note $A^\omega_{\sigma,u\ell} T^\omega_{\sigma,\ell} = A^\sigma_{\omega,u\ell} T^\sigma_{\omega,\ell}$)

$$\mathscr{T}[\widetilde{\mathfrak{N}}] - \mathscr{T}[\mathfrak{N}] = 2\varepsilon \sum_{\sigma,\omega=1}^n \sum_{\ell=1}^2 \iint_B A^\omega_{\sigma,u\ell} T^\omega_{\sigma,\ell} \, du\,dv + o(\varepsilon)$$

$$= 4\varepsilon \sum_{1\le\sigma<\omega\le n} \iint_B \left\{ A^\omega_{\sigma,u} T^\omega_{\sigma,1} + A^\omega_{\sigma,v} T^\omega_{\sigma,2} \right\} + o(\varepsilon)$$

$$= 4\varepsilon \sum_{1\le\sigma<\omega\le n} \int_{\partial B} A^\omega_\sigma (T^\omega_{\sigma,1}, T^\omega_{\sigma,2}) \cdot \nu \, ds$$

$$- 4\varepsilon \sum_{1\le\sigma<\omega\le n} \iint_B A^\omega_\sigma \operatorname{div} (T^\omega_{\sigma,1}, T^\omega_{\sigma,2}) \, du\,dv + o(\varepsilon).$$

But $\mathbf{A}(w)$ was chosen arbitrarily which proves the statement. $\square$

4.2.3 The Integral Functions

Interpreting the Euler–Lagrange equations as integrability conditions, analogously to the situation considered in Sect. 3.4.2, we find integral functions

$$\tau^{(\sigma\vartheta)} \in C^{k-1}(\overline{B}, \mathbb{R})$$

satisfying

$$\nabla \tau^{(\sigma\vartheta)} = \left(- T^\vartheta_{\sigma,2}, T^\vartheta_{\sigma,1} \right) \quad \text{in } B \quad \text{for all } \sigma, \vartheta = 1, \ldots, n,$$

using again conformal parameters. The boundary conditions $(T_{\sigma,1}^{\vartheta}, T_{\sigma,2}^{\vartheta}) \cdot v = 0$ now imply

$$\nabla \tau^{(\sigma\vartheta)} \cdot (-v, u) = 0 \quad \text{on } \partial B$$

with the unit tangent vector $(-v, u)$ at ∂B, and we may again choose $\tau^{(\sigma\vartheta)}$ so that

$$\tau^{(\sigma\vartheta)} = 0 \quad \text{on } \partial B \quad \text{for all } \sigma, \vartheta = 1, \ldots, n.$$

Note that the matrix $(\tau^{(\sigma\vartheta)})_{\sigma,\vartheta=1,\ldots,n}$ is skew-symmetric.

4.2.4 A Non-linear Elliptic System

With the above integral functions we now define the quantities

$$\delta \tau^{(\sigma\vartheta)} := \sum_{\omega=1}^{n} \det\left(\nabla \tau^{(\sigma\omega)}, \nabla \tau^{(\omega\vartheta)}\right), \quad \sigma, \vartheta = 1, \ldots n.$$

The matrix $(\delta \tau^{(\sigma\vartheta)})_{\sigma,\vartheta=1,\ldots,n}$ is also skew-symmetric. The special nature of the $\delta \tau^{(\sigma\vartheta)}$ will be worked out in following paragraphs.

The aim in this paragraph is to establish an elliptic system with quadratic growth in the gradient of $\tau^{(\sigma\vartheta)}$ for normal Coulomb frames.

Proposition 4.2. *Let a normal Coulomb frame $\mathfrak{N}$ for the conformally parametrized immersion $X \colon B \to \mathbb{R}^{n+2}$ be given. Then the integral functions $\tau^{(\sigma\vartheta)}$ are solutions of the boundary value problems*

$$\Delta \tau^{(\sigma\vartheta)} = -\delta \tau^{(\sigma\vartheta)} + S_{\sigma,12}^{\vartheta} \quad \text{in } B, \quad \tau^{(\sigma\vartheta)} = 0 \quad \text{on } \partial B,$$

where $\delta \tau^{(\sigma\vartheta)}$ grows quadratically in the gradient $\nabla \tau^{(\sigma\vartheta)}$.

Proof. Choose $(\sigma, \vartheta) \in \{1, \ldots, n\} \times \{1, \ldots, n\}$ and recall the identity

$$S_{\sigma,12}^{\vartheta} = \partial_v T_{\sigma,1}^{\vartheta} - \partial_u T_{\sigma,2}^{\vartheta} + \sum_{\omega=1}^{n} \left\{ T_{\sigma,1}^{\omega} T_{\omega,2}^{\vartheta} - T_{\sigma,2}^{\omega} T_{\omega,1}^{\vartheta} \right\}$$

for the components of the normal curvature tensor. Together with $\nabla \tau^{(\sigma\vartheta)} = (-T_{\sigma,2}^{\vartheta}, T_{\sigma,1}^{\vartheta})$ we arrive at

$$\Delta \tau^{(\sigma\vartheta)} = \partial_v T_{\sigma,1}^{\vartheta} - \partial_u T_{\sigma,2}^{\vartheta} = -\sum_{\omega=1}^{n} \left\{ T_{\sigma,1}^{\omega} T_{\omega,2}^{\vartheta} - T_{\sigma,2}^{\omega} T_{\omega,1}^{\vartheta} \right\} + S_{\sigma,12}^{\vartheta}$$

$$= \sum_{\omega=1}^{n} \left\{ \tau_v^{(\sigma\omega)} \tau_u^{(\omega\vartheta)} - \tau_u^{(\sigma\omega)} \tau_v^{(\omega\vartheta)} \right\} + S_{\sigma,12}^{\vartheta}$$

proving the statement. $\qquad\square$

4.3   Examples

4.3.1   The Case $n = 2$

In this case there is only one integral function $\tau^{(12)}$ satisfying

$$\Delta\tau^{(12)} = S^2_{1,12} \quad \text{in } B, \quad \tau^{(12)} = 0 \quad \text{on } \partial B.$$

This is exactly the Poisson equation with homogeneous boundary data from Sect. 3.4.2 where we used the notations $\tau = \tau^{(12)}$ and $SW = S^2_{1,12}$.

4.3.2   The Case $n = 3$

If $n = 3$ we have three relations

$$\Delta\tau^{(12)} = \tau_v^{(13)}\tau_u^{(32)} - \tau_u^{(13)}\tau_v^{(32)} + S^2_{1,12},$$
$$\Delta\tau^{(13)} = \tau_v^{(12)}\tau_u^{(23)} - \tau_u^{(12)}\tau_v^{(23)} + S^3_{1,12},$$
$$\Delta\tau^{(23)} = \tau_v^{(21)}\tau_u^{(13)} - \tau_u^{(21)}\tau_v^{(13)} + S^3_{2,12}.$$

Proposition 4.3. *Let a conformally parametrized immersion* $X \colon \overline{B} \to \mathbb{R}^5$ *together with a normal Coulomb frame* $\mathfrak{N}$ *be given. Let*

$$\mathfrak{T} := (\tau^{(12)}, \tau^{(13)}, \tau^{(23)}), \quad \text{and recall} \quad \mathfrak{S} = \frac{1}{W}\left(S^2_{1,12}, S^3_{1,12}, S^3_{2,12}\right)$$

with the normal curvature vector $\mathfrak{S}$ *from Sect. 1.6.7. Then it holds*

$$\Delta\mathfrak{T} = \mathfrak{T}_u \times \mathfrak{T}_v + \mathfrak{S}W \quad \text{in } B, \quad \mathfrak{T} = 0 \quad \text{on } \partial B$$

with the usual vector product $\times$ *in* $\mathbb{R}^3$. *Thus, the vector* $\mathfrak{T}$ *solves an inhomogeneous H-surface system with constant mean curvature* $H = \frac{1}{2}$ *and vanishing boundary data.*

Namely, compare it with the mean curvature system in $\mathbb{R}^3$ from Sect. 2.1.3, i.e.

$$\Delta X = 2HWN \quad \text{in } B.$$

If X additionally satisfies the conformality relations, then it would actually represent an immersion with scalar mean curvature H.

Let us come back to the above differential system. From Wente [122] we infer[1]

---

[1] See also Sect. 4.5.1, and consult Sect. 4.5.3 for a new proof of Wente's result.

Proposition 4.4. *Let the immersion* $X : \overline{B} \to \mathbb{R}^5$ *with flat normal bundle* $\mathfrak{S} \equiv 0$ *be given. Then it holds* $\mathfrak{T} \equiv 0$.

In other words: *A normal Coulomb frame for an immersion in* $\mathbb{R}^5$ *is parallel* - insofar it exists at all. The missing existence proof together with classical regularity results will be discussed later.

4.4   Quadratic Growth in the Gradient

4.4.1   A Grassmann-Type Vector

The latter example gives rise to a final definition of the following vector of Grassmann type

$$\mathfrak{T} := \left(\tau^{(\sigma\vartheta)}\right)_{1 \le \sigma < \vartheta \le n} \in \mathbb{R}^N, \quad N := \frac{n}{2}\,(n-1).$$

In our examples above, this vector $\mathfrak{T}$ works as follows:

$$\begin{aligned}
\mathfrak{T} &= \tau^{(12)} \in \mathbb{R} && \text{for } n = 2, \\
\mathfrak{T} &= \left(\tau^{(12)}, \tau^{(13)}, \tau^{(23)}\right) \in \mathbb{R}^3 && \text{for } n = 3.
\end{aligned}$$

Analogously we define the Grassmann-type vector

$$\delta\mathfrak{T} := \left(\delta\tau^{(\sigma\vartheta)}\right)_{1 \le \sigma < \vartheta \le n} \in \mathbb{R}^N.$$

Then the proposition from Sect. 4.2.4 can be rewritten in the form

$$\Delta\mathfrak{T} = -\delta\mathfrak{T} + \mathfrak{S}W \quad \text{in } B, \quad \mathfrak{T} = 0 \quad \text{on } \partial B.$$

From the definition of $\delta\mathfrak{T}$ we immediately obtain

$$|\Delta\mathfrak{T}| \le c\,|\nabla\mathfrak{T}|^2 + |\mathfrak{S}W| \quad \text{in } B$$

with a suitable real constant $c > 0$.

4.4.2   A Non-linear System with Quadratic Growth

The exact knowledge of this constant $c > 0$ will become important later.

Proposition 4.5. *Using conformal parameters* $(u, v) \in \overline{B}$, *it holds*

$$|\Delta\mathfrak{T}| \le \frac{\sqrt{n-2}}{2}|\nabla\mathfrak{T}|^2 + |\mathfrak{S}W| \quad in \; B.$$

Proof. We already know

$$|\Delta\mathfrak{T}| \le |\delta\mathfrak{T}| + |\mathfrak{S}W| \quad in \; B.$$

It remains to estimate $|\delta\mathfrak{T}|$ appropriately. We begin with computing

$$|\delta\mathfrak{T}|^2 = \sum_{1 \le \sigma < \vartheta \le n} \left\{ \sum_{\omega=1}^{n} \det\left(\nabla\tau^{(\sigma\omega)}, \nabla\tau^{(\omega\vartheta)}\right) \right\}^2$$

$$\le (n-2) \sum_{1 \le \sigma < \vartheta \le n} \left\{ \sum_{\omega=1}^{n} \det\left(\nabla\tau^{(\sigma\omega)}, \nabla\tau^{(\omega\vartheta)}\right)^2 \right\}.$$

Note that only derivatives of elements of $\mathfrak{T}$ appear on the right hand side of $|\delta\mathfrak{T}|^2$. Moreover, this right hand side can be estimated by $|\mathfrak{T}_u \wedge \mathfrak{T}_v|^2$ since $\mathfrak{T}_u \wedge \mathfrak{T}_v$ has actually more elements.[2] Thus using Lagrange's identity

$$|X \wedge Y|^2 = |X|^2|Y|^2 - (X \cdot Y)^2 \le |X|^2|Y|^2$$

we can estimate as follows

$$|\delta\mathfrak{T}|^2 \le (n-2)|\mathfrak{T}_u \wedge \mathfrak{T}_v|^2 \le (n-2)|\mathfrak{T}_u|^2|\mathfrak{T}_v|^2.$$

Taking all together shows

$$|\Delta\mathfrak{T}| \le \sqrt{n-2}\,|\mathfrak{T}_u||\mathfrak{T}_v| + |\mathfrak{S}W| \le \frac{\sqrt{n-2}}{2}|\nabla\mathfrak{T}|^2 + |\mathfrak{S}W|$$

which proves the statement.   □

In the special case $n = 2$ we immediately verifyf

$$|\Delta\tau| \le |SW| \quad in \; B$$

with the scalar normal curvature $\mathfrak{S} = S$.

---

[2] In particular, elements of the form $\det\left(\nabla\tau^{(\sigma\omega)}, \nabla\tau^{(\omega'\vartheta)}\right)^2$ appear in $|\mathfrak{T}_u \wedge \mathfrak{T}_v|^2$, but they do not appear in the right hand side of the inequality.

4.5 Torsion-Free Normal Frames

4.5.1 The Case $n = 3$

As we mentioned above, the only solution (with finite Dirichlet energy) of the elliptic system

$$\Delta \mathfrak{T} = \mathfrak{T}_u \times \mathfrak{T}_v \quad \text{in } B, \quad \mathfrak{T} = 0 \quad \text{on } \partial B,$$

is $\mathfrak{T} \equiv 0$, see Wente [122]. This is exactly the situation for conformally parametrized immersions $X : \overline{B} \to \mathbb{R}^5$ with flat normal bundle $\mathfrak{S} = 0$.

A proof of Wente's result would follow from asymptotic expansions of the solutions $\mathfrak{T}$ in the interior B and on the boundary ∂B as we will employ in Sect. 4.5.4. For such asymptotic expansions we particularly refer to Hartman and Wintner [59], Heinz [63] or Hildebrandt [66]. But a new and simple proof is presented as a corollary in paragraph Sect. 4.5.3.

Corollary 4.1. *Suppose that the immersion* $X : \overline{B} \to \mathbb{R}^5$ *admits a normal Coulomb frame. Then this frame is free of torsion if and only if the curvature vector* $\mathfrak{S}$ *vanishes identically.*

This is a special case of a general result we will discuss in Sect. 4.5.4.

4.5.2 An Auxiliary Function and a New Proof of Wente's Result

To handle the general case $n \geq 3$ we need some preparations. Let us start with the

Lemma 4.1. *Let the conformally parametrized immersion* $X : \overline{B} \to \mathbb{R}^{n+2}$ *together with a normal Coulomb frame* $\mathfrak{N}$ *be given. Assume* $\mathfrak{S} = 0$. *Then the function*

$$\Phi(w) = \mathfrak{T}_w(w) \cdot \mathfrak{T}_w(w) = \sum_{1 \leq \sigma < \vartheta \leq n} \tau_w^{(\sigma \vartheta)} \tau_w^{(\sigma \vartheta)}$$

with the complex derivative $\phi_w = \frac{1}{2}(\phi_u - i\phi_v)$ *vanishes identically in* $\overline{B}$.

Proof. We will prove that Φ solves the boundary value problem

$$\Phi_{\overline{w}} = 0 \quad \text{in } B, \quad \text{Im}(w^2 \Phi) = 0 \quad \text{on } \partial B.$$

Then the analytic function $\Psi(w) := w^2 \Phi(w)$ has vanishing imaginary part, and the Cauchy–Riemann equations imply $\Psi(w) \equiv c \in \mathbb{R}$. The assertion follows then finally from $\Psi(0) = 0$.

1. In order to deduce the stated boundary condition, recall that $\tau^{(\sigma\vartheta)} = 0$ on ∂B. Thus all tangential derivatives vanish identically because

$$-v\tau_u^{(\sigma\vartheta)} + u\tau_v^{(\sigma\vartheta)} = -\operatorname{Im}(w\tau_w^{(\sigma\vartheta)}) = 0 \quad \text{on } \partial B$$

for all $\sigma, \vartheta = 1, \ldots, n$.

The boundary condition now follows from

$$\operatorname{Im}\left(w^2 \Phi\right) = \operatorname{Im}\left(w^2\, \mathfrak{T}_w \cdot \mathfrak{T}_w\right) = \operatorname{Im}\left\{w^2 \sum_{1 \le \sigma < \vartheta \le n} \tau_w^{(\sigma\vartheta)} \tau_w^{(\sigma\vartheta)}\right\}$$

$$= \sum_{1 \le \sigma < \vartheta \le n} \operatorname{Im}\left\{\left(w\tau_w^{(\sigma\vartheta)}\right)\left(w\tau_w^{(\sigma\vartheta)}\right)\right\}$$

$$= 2 \sum_{1 \le \sigma < \vartheta \le n} \operatorname{Re}\left(w\tau_w^{(\sigma\vartheta)}\right) \operatorname{Im}\left(w\tau_w^{(\sigma\vartheta)}\right) = 0$$

on the boundary ∂B.

2. Now we show the analyticity of Φ with the aid of the identities

$$\Delta\tau^{(\sigma\vartheta)} = 4\tau_{w\bar{w}}^{(\sigma\vartheta)} = -\delta\tau^{(\sigma\vartheta)}.$$

Interchanging indices cyclically yields

$$2\Phi_{\bar{w}} = 4\mathfrak{T}_w \cdot \mathfrak{T}_{w\bar{w}} = 4 \sum_{1 \le \sigma < \vartheta \le n} \tau_w^{(\sigma\vartheta)} \tau_{w\bar{w}}^{(\sigma\vartheta)} = \frac{1}{2} \sum_{\sigma,\vartheta=1}^{n} \tau_w^{(\sigma\vartheta)} \Delta\tau^{(\sigma\vartheta)}$$

$$= \frac{1}{4} \sum_{\sigma,\vartheta,\omega=1}^{n} \left\{\tau_v^{(\sigma\omega)} \tau_u^{(\omega\vartheta)} \tau_u^{(\sigma\vartheta)} - \tau_u^{(\sigma\omega)} \tau_v^{(\omega\vartheta)} \tau_u^{(\sigma\vartheta)}\right\}$$

$$- \frac{i}{4} \sum_{\sigma,\vartheta,\omega=1}^{n} \left\{\tau_v^{(\sigma\omega)} \tau_u^{(\omega\vartheta)} \tau_v^{(\sigma\vartheta)} - \tau_u^{(\sigma\omega)} \tau_v^{(\omega\vartheta)} \tau_v^{(\sigma\vartheta)}\right\}$$

$$= \frac{1}{4} \sum_{\sigma,\vartheta,\omega=1}^{n} \left\{\tau_v^{(\omega\vartheta)} \tau_u^{(\vartheta\sigma)} \tau_u^{(\omega\sigma)} - \tau_u^{(\sigma\omega)} \tau_v^{(\omega\vartheta)} \tau_u^{(\sigma\vartheta)}\right\}$$

$$- \frac{i}{4} \sum_{\sigma,\vartheta,\omega=1}^{n} \left\{\tau_v^{(\vartheta\sigma)} \tau_u^{(\sigma\omega)} \tau_v^{(\vartheta\omega)} - \tau_u^{(\sigma\omega)} \tau_v^{(\omega\vartheta)} \tau_v^{(\sigma\vartheta)}\right\}$$

which shows $\Phi_{\bar{w}} = 0$. The proof is complete.   □

This lemma gives us an interesting geometric interpretation of the present situation: Namely, there hold

$$\mathfrak{T}_w \cdot \mathfrak{T}_w = 0, \quad \text{or equivalently}$$

$$\mathfrak{T}_u \cdot \mathfrak{T}_u = \mathfrak{T}_v \cdot \mathfrak{T}_v, \quad \mathfrak{T}_u \cdot \mathfrak{T}_v = 0 \quad \text{in } B.$$

This means, in other words, that *in case* $\mathfrak{S} \equiv 0$ *of flat normal bundle the mapping* $\mathfrak{T}$ *fulfills the conformality relations.*

Does this observation implies further consequences? However, it is the key for our next considerations.

4.5.3 A New Proof of Wente's Result

But before the following remark is due: As Frank Müller pointed out in his lecture notes [90], the above proof provides a generalization of Wente's result from [122].

Corollary 4.2. *Let the immersion X solve the mean curvature system*

$$\Delta X = 2HWN = 2HX_u \times X_v \quad \text{in } B$$

in case $n = 1$ of one codimension with vanishing boundary data $X = 0$ on ∂B and scalar mean curvature $H \in C^0(B, \mathbb{R})$. Then it holds $X \equiv 0$ in B.

Sketch of the proof. Let again

$$\Phi := X_w \cdot X_w = \frac{1}{4}(X_u^2 - X_v^2) - \frac{i}{2}X_u \cdot X_v.$$

Then we verify

$$\Phi_{\overline{w}} = \frac{1}{4}X_u \cdot \Delta X - \frac{i}{4}X_v \cdot \Delta X = 0 \quad \text{in } B$$

as well as

$$\text{Im}\,(wX_w) = \text{Im}\,(e^{i\varphi}X_w(e^{i\varphi})) = -\frac{1}{2}\frac{\partial}{\partial\varphi}X(e^{i\varphi}) = 0 \quad \text{on } \partial B.$$

As in the proof of the previous lemma, we infer

$$\text{Im}\,(w^2\Phi) = 0 \quad \text{on } \partial B,$$

and the statement follows as above.   □

Note that the same is true for immersions with prescribed mean curvature vector $\mathfrak{H}$ in the general case $n \geq 1$ if we additionally assume the conformality relations for X, since then it holds

$$\Delta X = 2 \sum_{\vartheta=1}^{n} H_\vartheta W N_\vartheta$$

as we know from Sect. 2.1.3. Thus, X_u and X_v would again be orthogonal to ΔX as needed in the proof. But the point is that Wente's result is independent of such a parametrization.

4.5.4 The Case $n > 3$

The main result of this section is the

Theorem 4.1. *Suppose that the immersion $X : \overline{B} \rightarrow \mathbb{R}^{n+2}$ admits a normal Coulomb frame $\mathfrak{N}$. Then this frame is free of torsion if and only if the curvature vector $\mathfrak{S}$ of its normal bundle vanishes identically.*

Proof. Let $\mathfrak{N}$ be a normal Coulomb frame. If it is free of torsion then $\mathfrak{S}$ vanishes identically. So assume conversely $\mathfrak{S} \equiv 0$ and let us show that $\mathfrak{N}$ is free of torsion. Consider for this purpose the Grassmann-type vector $\mathfrak{T}$ from above. Because it holds $\mathfrak{T}_w \cdot \mathfrak{T}_w \equiv 0$ by the previous lemma, we find (now independently of n)

$$|\mathfrak{T}_u| = |\mathfrak{T}_v|, \quad \mathfrak{T}_u \cdot \mathfrak{T}_v = 0 \quad \text{in } B,$$

i.e. $\mathfrak{T}$ is a conformally parametrized solution of

$$\Delta \mathfrak{T} = -\delta \mathfrak{T} \quad \text{in } B, \quad \mathfrak{T} = 0 \quad \text{on } \partial B.$$

According to the quadratic growth condition

$$|\delta \mathfrak{T}| \le c |\nabla \mathfrak{T}|^2$$

from Sect. 4.4.1, the arguments in Heinz [63] apply[3]: Assume $\mathfrak{T} \not\equiv \text{const}$ in B. Then the asymptotic expansion stated in the Satz of Heinz [63] implies that boundary branch points $w_0 \in \partial B$ with $\mathfrak{T}_u(w_0) = \mathfrak{T}_v(w_0) = 0$ are isolated. But this contradicts our boundary condition $\mathfrak{T}|_{\partial B} = 0$. Thus, $\mathfrak{T} \equiv \text{const} = 0$ which implies $\tau^{(\sigma\vartheta)} \equiv 0$ for all $\sigma, \vartheta = 1, \ldots, n$, and the normal Coulomb frame is free of torsion. □

The existence of torsion-free resp. parallel ONF's for surfaces with flat normal bundle is well established in the literature, we refer e.g. to Chen [21]. With the previous and the following results we provide alternative methods.

---

[3]Let $X \in C^2(\overline{B}, R^n)$ solve the elliptic system $\Delta X = Hf(X, X_u, X_v)$ with $X_u^2 = X_v^2, X_u \cdot X_v = 0$, where $|f(x, p, q)| \le \mu(|x|)(|p|^2 + |q|^2)$. Then if $X_u(w_0) = 0$ for some $w_0 \in \partial B$, but $X_u \not\equiv 0$, the following asymptotic expansion $X_w(w_0) = a(w - w_0)^\ell + o(|w - w_0|^\ell)$ holds true for $w \rightarrow w_0$, where $a \in \mathbb{C} \setminus \{0\}$ with $a_1^2 + \ldots + a_n^2 = 0$, and $\ell \in \mathbb{N} \setminus \{0\}$.

4.6 Non-flat Normal Bundles

We mainly want to focus on surfaces with non-flat normal bundles, and with normal
Coulomb frames we introduce a concept which replaces parallelism in the normal
bundles if such bundles are curved. We begin with establishing possible upper
bounds for the torsion coefficients of normal Coulomb frames.

4.6.1 An Upper Bound via Wente's L^∞-Estimate

Let $\| \cdot \|_{L^p(B)}$ denote the Lebesgue L^p-norm. We want to prove the

Proposition 4.6. *Let $\mathfrak{N}$ be a normal Coulomb frame for the conformally parametrized immersion X. Then the Grassmann-type vector $\mathfrak{T}$ satisfies*

$$\|\mathfrak{T}\|_{L^\infty(B)} \le \frac{n-2}{2\pi} \|\nabla \mathfrak{T}\|_{L^2(B)}^2 + \frac{1}{4} \|\mathfrak{S}W\|_{L^\infty(B)}.$$

Proof. 1. For $1 \le \sigma < \vartheta \le n$, $\omega \in \{1,\dots,n\}$ with $\omega \notin \{\sigma, \vartheta\}$, and given integral
functions $\tau^{(\sigma\vartheta)}$ we define the functions $y^{(\sigma\vartheta\omega)}$ as the unique solutions of

$$\Delta y^{(\sigma\vartheta\omega)} = -\det\left(\nabla \tau^{(\sigma\omega)}, \nabla \tau^{(\omega\vartheta)}\right) \quad \text{in } B, \quad y^{(\sigma\vartheta\omega)} = 0 \quad \text{on } \partial B.$$

Wente's L^∞-estimate from Wente [123] together with Topping [116] then yields
the *optimal inequalities*[4]

$$\|y^{(\sigma\vartheta\omega)}\|_{L^\infty(B)} \le \frac{1}{4\pi}\left(\|\nabla \tau^{(\sigma\omega)}\|_{L^2(B)}^2 + \|\nabla \tau^{(\omega\vartheta)}\|_{L^2(B)}^2\right)$$

for all $1 \le \sigma < \vartheta \le n$, $\omega \notin \{\sigma, \vartheta\}$. In addition, we introduce the Grassmann-
type vector $\mathfrak{z} = (z^{(\sigma\vartheta)})_{1 \le \sigma < \vartheta \le n}$ as the unique solution of

$$\Delta \mathfrak{z} = \mathfrak{S}W \quad \text{in } B, \quad \mathfrak{z} = 0 \quad \text{on } \partial B.$$

We use Poisson's representation formula and estimate as follows (see e.g. [107])

$$|\mathfrak{z}(w)| = \left| \iint_B \Phi(\zeta;w)\mathfrak{S}(\zeta)W(\zeta)\,d\xi d\eta \right| \le \|\mathfrak{S}W\|_{L^\infty(B)} \iint_B \Phi(\zeta;w)\,d\xi d\eta$$

---

[4]In 1980 H. Wente proved: Let $\Phi \in C^0(\overline{B}, \mathbb{R}) \cap H^{1,2}(B, \mathbb{R})$ solve $\Delta\Phi = -(f_u g_v - f_v g_u)$ in B with $\Phi = 0$ on $\partial\Omega$ and $f, g \in H^{1,2}(B, \mathbb{R})$, then $\|\Phi\|_{L^\infty(B)} + \|\nabla\Phi\|_{L^2(B)} \le C\|\nabla f\|_{L^2(B)}\|\nabla g\|_{L^2(B)}$. Following Topping [116] we may set $\frac{C}{2} = \frac{1}{4\pi}$ after applying Hölder's
inequality.

setting $\zeta = (\xi, \eta)$, which leads us to

$$|\mathfrak{Z}(w)| \le \frac{1}{4} \|\mathfrak{S}W\|_{L^\infty(B)}$$

with the Green function $\phi(\zeta; w)$ for Δ in B; see e.g. Sect. 3.4.2 where we proved

$$\iint_B \Phi(\zeta; w) \, d\xi d\eta \le \frac{1}{4}.$$

2. Now recall that

$$\Delta \tau^{(\sigma\vartheta)} = -\sum_{\omega=1}^{n} \det\left(\nabla\tau^{(\sigma\omega)}, \nabla\tau^{(\omega\vartheta)}\right) + S^\vartheta_{\sigma,12} = \sum_{\omega=1}^{n} \Delta y^{(\sigma\vartheta\omega)} + \Delta z^{(\sigma\vartheta)}.$$

Taking account of the unique solvability of the above mentioned Dirichlet problems with vanishing boundary data, the maximum principle yields

$$\tau^{(\sigma\vartheta)} = \sum_{\omega \notin \{\sigma, \vartheta\}} y^{(\sigma\vartheta\omega)} + z^{(\sigma\vartheta)}, \quad 1 \le \sigma < \vartheta \le n.$$

Now we collect all the estimates obtained and get (we rearrange the summations and redefine some indices of the sums)

$$\|\mathfrak{T}\|_{L^\infty(B)} \le \sum_{1 \le \sigma < \vartheta \le n} \sum_{\omega \notin \{\sigma,\vartheta\}} \|y^{(\sigma\vartheta\omega)}\|_{L^\infty(B)} + \|\mathfrak{Z}\|_{L^\infty(B)}$$

$$\le \frac{1}{4\pi} \sum_{1 \le \sigma < \vartheta \le n} \sum_{\omega \notin \{\sigma,\vartheta\}} \left(\|\nabla\tau^{(\sigma\omega)}\|^2_{L^2(B)} + \|\nabla\tau^{(\omega\vartheta)}\|^2_{L^2(B)} \right)$$

$$+ \frac{1}{4} \|\mathfrak{S}W\|_{L^\infty(B)}$$

$$= \frac{1}{4\pi} \left\{ \sum_{1 \le \omega < \sigma < \vartheta \le n} \left(\|\nabla\tau^{(\sigma\omega)}\|^2_{L^2(B)} + \|\nabla\tau^{(\omega\vartheta)}\|^2_{L^2(B)} \right) \right.$$

$$+ \sum_{1 \le \sigma < \omega < \vartheta \le n} \left(\|\nabla\tau^{(\sigma\omega)}\|^2_{L^2(B)} + \|\nabla\tau^{(\omega\vartheta)}\|^2_{L^2(B)} \right)$$

$$\left. + \sum_{1 \le \sigma < \vartheta < \omega \le n} \left(\|\nabla\tau^{(\sigma\omega)}\|^2_{L^2(B)} + \|\nabla\tau^{(\vartheta\omega)}\|^2_{L^2(B)} \right) \right\}$$

$$+ \frac{1}{4} \|\mathfrak{S}W\|_{L^\infty(B)}$$

$$= \frac{1}{2\pi} \sum_{1 \le \sigma < \vartheta \le n} \sum_{\omega \notin \{\sigma, \vartheta\}} \|\nabla \tau^{(\sigma \vartheta)}\|^2_{L^2(B)} + \frac{1}{4} \|\mathfrak{S}W\|_{L^\infty(B)}$$

$$= \frac{n-2}{2\pi} \|\nabla \mathfrak{T}\|^2_{L^2(B)} + \frac{1}{4} \|\mathfrak{S}W\|_{L^\infty(B)}.$$

This proves the statement. □

The dependence on the total torsion $\|\nabla \mathfrak{T}\|^2_{L^2(B)}$ has an unpleasant effect for higher codimensions $n > 2$. We do not know how we could get rid of this.

4.6.2 An Estimate for the Torsion Coefficients

We are now in the position to prove our main result of this section.

Theorem 4.2. *Let $\mathfrak{N}$ be a normal Coulomb frame for the conformally parametrized immersion $X: \overline{B} \to \mathbb{R}^{n+2}$ with total torsion $\mathscr{T}[\mathfrak{N}]$ and given $\|\mathfrak{S}W\|_{L^\infty(B)}$. Assume that the smallness condition*

$$\frac{\sqrt{n-2}}{2} \left(\frac{n-2}{2\pi} \mathscr{T}[\mathfrak{N}] + \frac{1}{4} \|\mathfrak{S}W\|_{L^\infty(B)} \right) < 1$$

is satisfied.
Then the torsion coefficients of $\mathfrak{N}$ can be estimated by

$$\|T^\vartheta_{\sigma,i}\|_{L^\infty(B)} \le c, \quad i = 1, 2, \ 1 \le \sigma < \vartheta \le n,$$

with a non-negative constant $c = c(n, \|\mathfrak{S}W\|_{L^\infty(B)}, \mathscr{T}[\mathfrak{N}]) < +\infty$.

Proof. From Sects. 4.4.2 and 4.6.1 we have the following elliptic system

$$|\Delta \mathfrak{T}| \le \frac{\sqrt{n-2}}{2} |\nabla \mathfrak{T}|^2 + |\mathfrak{S}W| \quad \text{in } B, \quad \mathfrak{T} = 0 \quad \text{on } \partial B,$$

$$\|\mathfrak{T}\|_{L^\infty(B)} \le \frac{n-2}{2\pi} \|\nabla \mathfrak{T}\|^2_{L^2(B)} + \frac{1}{4} \|\mathfrak{S}W\|_{L^\infty(B)} \le M \in [0, +\infty).$$

The smallness condition ensures that we can apply Heinz's global gradient estimate from Theorem 1 in [107], Chap. XII, Sect. 3, obtaining $\|\nabla \mathfrak{T}\|_\infty \le c$.[5] This in turn yields the desired estimate. □

---

[5] This theorem states: Let $X \in C^2(\overline{B}, \mathbb{R}^{n+2})$ be a solution of the elliptic system $|\Delta X| \le a|\nabla X|^2 + b$ in B with $X = 0$ on ∂B and $\|X\|_{L^\infty(B)} \le M$. Assume $aM < 1$. Then there is a constant $c = c(a, b, M, \alpha)$ such that $\|X\|_{C^{1+\alpha}(\overline{B})} \le c(a, b, M, \alpha)$.

It remains open to prove global pointwise estimates for the torsion coefficients without the smallness condition. As mentioned above, we would particularly like to get rid of the a priori dependence of $\mathscr{T}[\mathfrak{N}]$.

4.7 Bounds for the Total Torsion

4.7.1 Upper Bounds

From the torsion estimates above we can easily infer upper bounds for the functional of total torsion

$$\mathscr{T}[\mathfrak{N}] = 2 \sum_{1 \le \sigma < \vartheta \le n} \iint_B \left\{ (T_{\sigma,1}^{\vartheta})^2 + (T_{\sigma,2}^{\vartheta})^2 \right\} du\,dv$$

for normal Coulomb frames $\mathfrak{N}$. For example, the previous theorem yields an estimate of the form

$$\mathscr{T}[\mathfrak{N}] \le 2 \sum_{1 \le \sigma < \vartheta \le n} \iint_B \left\{ \|T_{\sigma,1}^{\vartheta}\|_{L^\infty(B)}^2 + \|T_{\sigma,2}^{\vartheta}\|_{L^\infty(B)}^2 \right\} du\,dv$$
$$=: C(n, \|\mathfrak{S}W\|_{L^\infty(B)}, \mathscr{T}[\mathfrak{N}])$$

but it remains to solve this inequality for $\mathscr{T}[\mathfrak{N}]$. Thus, let us mention briefly a second way which works for *small solutions* $\mathfrak{T}$:
Multiplying

$$\Delta \mathfrak{T} = -\delta \mathfrak{T} + \mathfrak{S}W$$

by $\mathfrak{T}$ and integrating by parts yields

Proposition 4.7. *For small solutions* $\|\mathfrak{T}\|_{L^\infty(B)} < \frac{2}{\sqrt{n-2}}$ *it holds*

$$\mathscr{T}[\mathfrak{N}] = 2\|\nabla \mathfrak{T}\|_{L^2(B)}^2 \le \frac{4\|\mathfrak{T}\|_{L^\infty(B)} \|\mathfrak{S}W\|_{L^1(B)}}{2 - \sqrt{n-2}\|\mathfrak{T}\|_{L^\infty(B)}}.$$

The reader is referred to Sauvigny [107] where such small solutions of non-linear elliptic systems are constructed. Let us emphasize that the case $n = 2$ is much easier to handle: The classical maximum principle controls $\|\mathfrak{T}\|_{L^\infty(B)}$ in terms of $\|\mathfrak{S}W\|_{L^\infty(B)}$, and no smallness condition is needed to bound the functional of total torsion.

4.7.2 A Lower Bound

We want to establish a lower bound for the functional of total torsion.

Proposition 4.8. *Let $\mathfrak{N}$ be a normal Coulomb frame for the conformally parametrized immersion X. Assume that the curvature vector of its normal bundle satisfies $\mathfrak{S} \neq 0$ and $\|\nabla\mathfrak{S}\|_{L^2(B)} > 0$. Then it holds*

$$\mathscr{T}[\mathfrak{N}] \geq \left(\sqrt{n-2}\, \|\mathfrak{S}\|_{L^\infty(B)} + \frac{\|\mathfrak{S}\|^2_{L^2(B)}}{(1-\rho)^2 \mathscr{S}_2[X;B_\rho]} + \frac{2\|\nabla\mathfrak{S}\|^2_{L^2(B)}}{\mathscr{S}_2[X;B_\rho]} \right)^{-1} \mathscr{S}_2[X;B_\rho]$$

with the functional

$$\mathscr{S}_2[X;B_\rho] := \iint\limits_{B_\rho} |\mathfrak{S}|^2 W \, du\, dv,$$

and where the radius $\rho \in (0,1)$ is chosen such that $\mathscr{S}_2[X;B_\rho] > 0$.

Proof. 1. Because of $\mathfrak{S} \neq 0$ there exists a $\rho = \rho(\mathfrak{S}) \in (0,1)$ with the property $\mathscr{S}_2[X;B_{\rho(\mathfrak{S})}] > 0$, and this $\rho = \rho(\mathfrak{S})$ serves as the radius ρ from our assumption. Now we choose a test function $\eta \in C^0(B,\mathbb{R}) \cap H_0^{1,2}(B,\mathbb{R})$ with the properties

$$\eta \in [0,1] \quad \text{in } B, \quad \eta = 1 \quad \text{in } B_\rho, \quad |\nabla\eta| \leq \frac{1}{1-\rho} \quad \text{in } B.$$

Multiplying $\Delta\mathfrak{T} = -\delta\mathfrak{T} + \mathfrak{S}W$ by $(\eta\mathfrak{S})$ and integrating by parts yields

$$\iint\limits_B \nabla\mathfrak{T}\cdot\nabla(\eta\mathfrak{S})\,du\,dv = \iint\limits_B \eta\,\delta\mathfrak{T}\cdot\mathfrak{S}\,du\,dv - \iint\limits_B \eta\,|\mathfrak{S}|^2 W\,du\,dv.$$

Taking

$$|\delta\mathfrak{T}| \leq \sqrt{n-2}\,|\mathfrak{T}_u||\mathfrak{T}_v| \leq \frac{\sqrt{n-2}}{2}|\nabla\mathfrak{T}|^2$$

from Sect. 4.4.2 into account, we estimate as follows

$$\iint\limits_{B_\rho} |\mathfrak{S}|^2 W\,du\,dv \leq \iint\limits_B \eta\,|\mathfrak{S}|^2 W\,du\,dv$$

$$\leq \iint\limits_B \eta\,|\delta\mathfrak{T}\cdot\mathfrak{S}|\,du\,dv + \iint\limits_B |\nabla\mathfrak{T}\cdot\nabla(\eta\mathfrak{S})|\,du\,dv$$

$$\leq \|\mathfrak{S}\|_{L^\infty(B)} \iint\limits_B \eta\,|\delta\mathfrak{T}|\,du\,dv + \iint\limits_B |\nabla\eta|\,|\mathfrak{S}|\,|\nabla\mathfrak{T}|\,du\,dv$$

$$+ \iint\limits_B \eta\,|\nabla\mathfrak{S}|\,|\nabla\mathfrak{T}|\,du\,dv$$

$$\leq \frac{\sqrt{n-2}}{2}\|\mathfrak{S}\|_{L^\infty(B)}\iint\limits_B|\nabla\mathfrak{T}|^2\,du\,dv$$

$$+\frac{\varepsilon}{2}\iint\limits_B|\mathfrak{S}|^2\,du\,dv+\frac{1}{2\varepsilon(1-\rho)^2}\iint\limits_B|\nabla\mathfrak{T}|^2\,du\,dv$$

$$+\frac{\delta}{2}\iint\limits_B|\nabla\mathfrak{S}|^2\,du\,dv+\frac{1}{2\delta}\iint\limits_B|\nabla\mathfrak{T}|^2\,du\,dv$$

with arbitrary real numbers $\varepsilon,\delta>0$. Summarizing we arrive at

$$\mathscr{S}_2[X;B_\rho]\leq\left(\frac{\sqrt{n-2}}{2}\|\mathfrak{S}\|_{L^\infty(B)}+\frac{1}{2\varepsilon(1-\rho)^2}+\frac{1}{2\delta}\right)\|\nabla\mathfrak{T}\|^2_{L^2(B)}$$

$$+\frac{\varepsilon}{2}\|\mathfrak{S}\|^2_{L^2(B)}+\frac{\delta}{2}\|\nabla\mathfrak{S}\|^2_{L^2(B)}.$$

2. Now we choose $\varepsilon:=\frac{1}{\|\mathfrak{S}\|^2_{L^2(B)}}\mathscr{S}_2[X;B_\rho]>0$. Rearranging for $\mathscr{S}_2[X;B_\rho]$ gives

$$\mathscr{S}_2[X;B_\rho]\leq\left(\sqrt{n-2}\,\|\mathfrak{S}\|_{L^\infty(B)}+\frac{\|\mathfrak{S}\|^2_{L^2(B)}}{(1-\rho)^2\mathscr{S}_2[X;B_\rho]}+\frac{1}{\delta}\right)\|\nabla\mathfrak{T}\|^2_{L^2(B)}$$

$$+\delta\|\nabla\mathfrak{S}\|^2_{L^2(B)}.$$

And since $\|\nabla\mathfrak{S}\|_{L^2(B)}>0$ we can insert $\delta:=\frac{1}{2}\|\nabla\mathfrak{S}\|^{-2}_{L^2(B)}\mathscr{S}_2[X;B_\rho]$ to get

$$\mathscr{S}_2[X;B_\rho]\leq2\Lambda\|\nabla\mathfrak{T}\|^2_{L^2(B)}$$

$$\text{setting}\quad\Lambda:=\sqrt{n-2}\,\|\mathfrak{S}\|_{L^\infty(B)}+\frac{\|\mathfrak{S}\|^2_{L^2(B)}}{(1-\rho)^2\mathscr{S}[X;B_\rho]}+\frac{2\|\nabla\mathfrak{S}\|^2_{L^2(B)}}{\mathscr{S}_2[X;B_\rho]}.$$

Together with $\mathscr{T}[\mathfrak{N}]=2\|\nabla\mathfrak{T}\|^2_{L^2(B)}$ we arrive at the stated estimate.   □

4.8 Existence and Regularity of Weak Normal Coulomb Frames

4.8.1 The Dirichlet Problem for the Poisson Equation

To introduce the function spaces coming next into play we consider the Dirichlet boundary value problem

$$\Delta\phi(u, v) = r(u, v) \quad \text{in } B, \quad \phi(u, v) = 0 \quad \text{on } \partial B. \tag{DP}$$

Let us abbreviate this problem by (DP).

4.8.2 Schauder Estimates

Assume that $r \in C^\alpha(\overline{B}, \mathbb{R})$ holds true for the right hand side r. The classical potential theory shows that there exits a solution ϕ of (DP) which satisfies the following estimates

$$\|\phi\|_{C^1(\overline{B})} \leq c_1 \|r\|_{C^0(\overline{B})}$$

or, if one wants to establish higher regularity,

$$\|\phi\|_{C^2(\overline{B})} \leq c_2(\alpha) \|r\|_{C^\alpha(\overline{B})}.$$

The constants $c_1, c_2 \in [0, \infty)$ depend on the domain $\overline{B}$, while c_2 additionally depends on the Hölder exponent $\alpha \in (0, 1)$. We refer e.g. to Gilbarg and Trudinger [53], Kalf, Kriecherbauer and Wienholtz [77], or Sauvigny [107].

4.8.3 L^p-Estimates

Now let $r \in L^2(B, \mathbb{R})$. Then a weak solution $\phi \in H^{1,2}(B, \mathbb{R})$ of (DP) is of class $H^{2,2}(B, \mathbb{R})$, and it holds the a priori estimate

$$\|\phi\|_{H^{2,2}(B)} \leq C \left(\|\phi\|_{H^{1,2}(B)} + \|r\|_{L^2(B)} \right).$$

In particular, if $r \in H^{m-2,2}(B, \mathbb{R})$ for the right hand side, we have higher regularity

$$\|\phi\|_{H^{m,2}(B)} \leq C \left(\|\phi\|_{H^{1,2}(B)} + \|r\|_{H^{m-2,2}(B)} \right).$$

For detailed considerations we refer the reader to Dobrowolski [37], Satz 7.4, Gilbarg and Trudinger [53], or Sauvigny [107].

Note that we must require $r \in L^2(B, \mathbb{R})$ to infer higher regularity $\phi \in C^0(B, \mathbb{R})$ because $H^{2,2}(B, \mathbb{R})$ is continuously embedded in $C^0(B, \mathbb{R})$ by Sobolev's theorem.

But if only, on the other hand, $r \in L^1(B, \mathbb{R})$, then a weak solution $\phi \in H^{1,2}(B, \mathbb{R})$ of (DP) satisfies a priori

$$\|\phi\|_{L^q(B)} \leq C \|r\|_{L^1(B)} \quad \text{for all } 1 \leq q < \infty,$$

$$\|\nabla\phi\|_{L^p(B)} \leq C \|r\|_{L^1(B)} \quad \text{for all } 1 \leq p < 2.$$

A function $\phi \in H^{1,2}(B, \mathbb{R})$ is not necessarily continuous, or, in other words, a right hand side $r \in L^1(B, \mathbb{R})$ is too weak for a good regularity theory for our purposes.

4.8.4   Wente's L^∞-Estimate

The situation changes dramatically if the right hand side r possesses a certain algebraic structure. Namely, assume that

$$r = \frac{\partial a}{\partial u} \frac{\partial b}{\partial v} - \frac{\partial a}{\partial v} \frac{\partial b}{\partial u} \qquad \text{(RHS)}$$

with functions $a, b \in H^{1,2}(B, \mathbb{R})$. Then it holds again $r \in L^1(B, \mathbb{R})$, *but a solution $\phi \in H^{1,2}(B, \mathbb{R})$ is actually of class $C^0(B, \mathbb{R})$ and satisfies Wente's L^∞-estimate*

$$\|\phi\|_{L^\infty(B)} + \|\nabla \phi\|_{L^2(B)} \leq \frac{1}{4\pi} \|\nabla a\|_{L^2(B)} \|\nabla b\|_{L^2(B)} ;$$

see Helein [64], or the original approach of Wente [123].

We already used this inequality in Sect. 4.6.1 for establishing an upper bound for the total torsion of a normal Coulomb frame.

The idea of the proof of Wente's inequality follows from an ingenious partial integration of the right hand side r which is actually given in curl-form, and using the conformal invariance of the differential equation. An approximation of a and b by smooth functions a_n, b_n forming a Cauchy sequence with continuous limit would complete the proof.

4.8.5   Hardy Spaces

Wente's discovery can be considered as the starting point of the modern harmonic analysis. Its general framework is the concept of *Hardy* and *Lorentz spaces*.

From Helein [64] we will quote two definitions of Hardy spaces leading to equivalent formulations of the theory. For profound and comprehensive presentations of the underlying ideas and methods of harmonic analysis we refer to Moser [89] and Stein [114]; see also the recent paper of Sharp and Topping [112]. Finally we refer to Duren [39] for the classical theory of Hardy spaces in complex analysis.

Definition 4.2. (Tempered-distribution definition)
Let $\Psi \in C_0^\infty(\mathbb{R}^m)$ such that

$$\int_{\mathbb{R}^m} \Psi(x) \, dx = 1.$$

For each $\varepsilon > 0$ we set

$$\Psi_\varepsilon(x) = \frac{1}{\varepsilon^m}\,\Psi(\varepsilon^{-1}x),$$

and for $\phi \in L^1(\mathbb{R}^m)$ we define

$$\phi^*(x) = \sup_{\varepsilon>0} |(\Psi_\varepsilon \star \phi)(x)|.$$

Then ϕ belongs to the Hardy space $\mathscr{H}^1(\mathbb{R}^m)$ if and only if $\phi^* \in L^1(\mathbb{R}^m)$ with norm

$$\|\phi\|_{\mathscr{H}^1(\mathbb{R}^m)} = \|\phi\|_{L^1(\mathbb{R}^m)} + \|\phi^*\|_{L^1(\mathbb{R}^m)}.$$

Definition 4.3. (Riesz–Fourier-transform definition)
For any function $\phi \in L^1(\mathbb{R}^m)$ we denote by $R_\alpha\phi$ the function defined by

$$\mathscr{F}(R_\alpha\phi) = \frac{\xi_\alpha}{|\xi|}\,\mathscr{F}(\phi)(\xi)$$

with the ϕ-Fourier transform

$$\mathscr{F}(\phi)(\xi) = \frac{1}{(2\pi)^{\frac{m}{2}}} \int_{\mathbb{R}^m} e^{-ix\cdot\xi}\phi(x)\,dx.$$

Then ϕ belongs to the Hardy space $\mathscr{H}^1(\mathbb{R}^m)$ if and only if

$$R_\alpha\phi \in L^1(\mathbb{R}^m) \quad \text{for all } \alpha = 1,\dots,m$$

with the norm

$$\|\phi\|_{\mathscr{H}^1(\mathbb{R}^m)} = \|\phi\|_{L^1(\mathbb{R}^m)} + \sum_{\alpha=1}^{m} \|R_\alpha\phi\|_{L^1(\mathbb{R}^m)}.$$

Consider again our Dirichlet problem (DP) with (RHS): Let $a, b \in H^{1,2}(B,\mathbb{R})$, and denote by $a \mapsto \widehat{a}$ and $b \mapsto \widehat{b}$ its extensions in $H^{1,2}$ to the whole space $\mathbb{R}^2$ such that these mappings are continuous from $H^{1,2}(B,\mathbb{R})$ to $H^{1,2}(\mathbb{R}^2,\mathbb{R})$. Then, referring again to Helein [64], it holds $r \in \mathscr{H}^1(\mathbb{R}^2)$.

4.8.6 Lorentz Interpolation Spaces

Thus, this latter fact becomes especially important when we consider solutions ϕ of (DP) together with the special right hand side (RHS).

Definition 4.4. Let $\Omega \subset \mathbb{R}^m$ be open, and let $p \in (1, +\infty)$ and $q \in [1, +\infty]$. The *Lorentz space* $L^{(p,q)}(\Omega)$ is the set of measurable functions $\phi \colon \Omega \to \mathbb{R}$ such that

$$\|f\|_{L^{(p,q)}} := \left(\int\limits_0^\infty \{t^{\frac{1}{p}} \phi^*(t)\}^q \, \frac{dt}{t} \right)^{\frac{1}{q}} < \infty \quad \text{if } q < +\infty$$

or $\|f\|_{L^{(p,q)}} := \sup\limits_{t>0} t^{\frac{1}{p}} \phi^*(t) < \infty$ if $q = +\infty$. Here ϕ^* denotes the unique non-increasing rearrangement of $|\phi|$ on $[0, \text{meas } \Omega]$.

Lorentz spaces are Banach spaces with a suitable norm. They may be considered as a deformation of L^p. Notice that (see [64], Sect. 3.3)

$$L^{(p,p)}(B) = L^p(B), \quad L^{(p,1)}(B) \subset L^{(p,q')}(B) \subset L^{(p,q'')}(B) \subset L^{(p,\infty)}(B)$$

for $1 < q' < q''$. Then:

(a) If $\phi \in H^{1,2}(B, \mathbb{R})$ solves (DP), (RHS) with $r \in \mathcal{H}^1(\mathbb{R}^2)$ then $\frac{\partial \phi}{\partial x}, \frac{\partial \phi}{\partial y} \in L^{(2,1)}(B)$.

(b) In this situation we have $\phi \in C^0(B, \mathbb{R})$.

4.8.7 The General Regularity Result

Summarizing we can state the following regularity result, taken from Helein [64], Chap. 3, Sect. 3.3.

Proposition 4.9. *Let $\phi \in H^{1,2}(B, \mathbb{R})$ solve (DP) with a right hand side r given as in (RHS) with $a, b \in H^{1,2}(B, \mathbb{R})$. Then $\frac{\partial \phi}{\partial u}, \frac{\partial \phi}{\partial v} \in L^{(2,1)}(B)$, and it particularly holds $\phi \in C^0(\overline{B}, \mathbb{R})$.*

4.8.8 Existence of Weak Normal Coulomb Frames

In case $n = 2$ we constructed critical points of the functional $\mathcal{T}[\mathfrak{N}]$ of total torsion by solving the Euler–Lagrange equation directly and verified their minimum character. Now, in the general situation, we construct critical points by means of direct methods of the calculus of variations. We start with the following

Definition 4.5. Let $m \in \mathbb{N}$, $m \geq 2$. For two matrices $\mathbf{A}, \mathbf{B} \in \mathbb{R}^{m \times m}$ we define their *inner product*

$$\langle \mathbf{A}, \mathbf{B} \rangle := \text{trace} \, (\mathbf{A} \circ \mathbf{B}^t) = \sum_{\sigma, \vartheta = 1}^m A_\sigma^\vartheta B_\sigma^\vartheta$$

as well as the associated norm

$$|\mathbf{A}| := \sqrt{\langle \mathbf{A}, \mathbf{A} \rangle} = \sqrt{\sum_{\sigma,\vartheta=1}^{m} (A_\sigma^\vartheta)^2} \, .$$

Helein proved in [64], Lemma 4.1.3, existence of weak Coulomb frames in the tangent bundle of surfaces. We want to carry out his arguments and adapt his methods to construct weak normal Coulomb frames in our situation. Additionally we want to present a method to establish classical regularity of such ONF's.

In the following we always use conformal parameters $(u, v) \in \overline{B}$.

Proposition 4.10. *There exists a weak normal Coulomb frame*

$$\mathfrak{N} \in H^{1,2}(B) \cap L^\infty(B)$$

minimizing the functional $\mathscr{T}[\mathfrak{N}]$ of total torsion in the set of all weak ONF's of class $H^{1,2}(B) \cap L^\infty(B)$.

Proof. We fix[6] some ONF $\widetilde{\mathfrak{N}} \in C^{k-1,\alpha}(\overline{B})$ and interpret $\mathscr{T}[\mathfrak{N}]$ as a functional $\mathscr{F}(\mathbf{R})$ of $SO(n)$-regular rotations

$$\mathbf{R} = (R_\sigma^\vartheta)_{\sigma,\vartheta=1,\dots,n} \in H^{1,2}(B, SO(n))$$

by setting

$$\mathscr{F}(\mathbf{R}) = \sum_{\sigma,\vartheta=1}^{n} \sum_{i=1}^{2} \iint_B (T_{\sigma,i}^\vartheta)^2 \, du\,dv = \iint_B \left(|\mathbf{T}_1|^2 + |\mathbf{T}_2|^2 \right) du\,dv$$

where $\mathbf{T}_i = (T_{\sigma,i}^\vartheta)_{\sigma,\vartheta=1,\dots,n}$ (see Sect. 1.6.5) and

$$N_\sigma := \sum_{\vartheta=1}^{n} R_\sigma^\vartheta \widetilde{N}_\vartheta \, , \quad \sigma = 1, \dots, n.$$

Choose a minimizing sequence

$${}^\ell\mathbf{R} = ({}^\ell R_\sigma^\vartheta)_{\sigma,\vartheta=1,\dots,n} \in H^{1,2}(B, SO(n))$$

and define

$${}^\ell N_\sigma := \sum_{\vartheta=1}^{n} {}^\ell R_\sigma^\vartheta \widetilde{N}_\vartheta \, .$$

---

[6]Note that now we start from $\widetilde{\mathfrak{N}}$ and transform into $\mathfrak{N}$.

As in Sect. 1.6.5 we find[7]

$$^{\ell}\mathbf{T}_i = {}^{\ell}\mathbf{R}_{u^i} \circ {}^{\ell}\mathbf{R}^t + {}^{\ell}\mathbf{R} \circ \widetilde{\mathbf{T}}_i \circ {}^{\ell}\mathbf{R}^t$$

which implies

$$^{\ell}\mathbf{T}_i \circ {}^{\ell}\mathbf{T}_i^t = ({}^{\ell}\mathbf{R}_{u^i} \circ {}^{\ell}\mathbf{R}^t + {}^{\ell}\mathbf{R} \circ \widetilde{\mathbf{T}}_i \circ {}^{\ell}\mathbf{R}^t) \circ ({}^{\ell}\mathbf{R} \circ {}^{\ell}\mathbf{R}_{u^i}^t + {}^{\ell}\mathbf{R} \circ \widetilde{\mathbf{T}}_i^t \circ {}^{\ell}\mathbf{R}^t)$$
$$= {}^{\ell}\mathbf{R}_{u^i} \circ {}^{\ell}\mathbf{R}_{u^i}^t + {}^{\ell}\mathbf{R} \circ \widetilde{\mathbf{T}}_i \circ {}^{\ell}\mathbf{R}_{u^i}^t + {}^{\ell}\mathbf{R}_{u^i} \circ \widetilde{\mathbf{T}}_i^t \circ {}^{\ell}\mathbf{R}^t + {}^{\ell}\mathbf{R} \circ \widetilde{\mathbf{T}}_i \circ \widetilde{\mathbf{T}}_i^t \circ {}^{\ell}\mathbf{R}^t.$$

In particular, we conclude

$$\text{trace}\,({}^{\ell}\mathbf{T}_i \circ {}^{\ell}\mathbf{T}_i^t) = \text{trace}\,({}^{\ell}\mathbf{R}_{u^i} \circ {}^{\ell}\mathbf{R}_{u^i}^t) + 2\,\text{trace}\,({}^{\ell}\mathbf{R} \circ \widetilde{\mathbf{T}}_i \circ {}^{\ell}\mathbf{R}_{u^i}^t) + \text{trace}\,(\widetilde{\mathbf{T}}_i \circ \widetilde{\mathbf{T}}_i^t),$$

or using our notion of a matrix norm

$$|^{\ell}\mathbf{T}_i|^2 = |^{\ell}\mathbf{R}_{u^i}|^2 + 2\langle {}^{\ell}\mathbf{R} \circ \widetilde{\mathbf{T}}_i, {}^{\ell}\mathbf{R}_{u^i}\rangle + |\widetilde{\mathbf{T}}_i|^2. \tag{4.1}$$

Furthermore, taking $|^{\ell}\mathbf{R} \circ \widetilde{\mathbf{T}}_i| = |\widetilde{\mathbf{T}}_i|$ into account, we arrive at the estimate

$$|^{\ell}\mathbf{T}_i|^2 \geq (|\widetilde{\mathbf{T}}_i| - |^{\ell}\mathbf{R}_{u^i}|)^2 \quad \text{a.e. on } B \quad \text{for all } \ell \in \mathbb{N}.$$

Now because the $\widetilde{\mathbf{T}}_i$ are bounded in $L^2(B, \mathbb{R}^{n \times n})$, and since $^{\ell}\mathbf{R}$ is minimizing for $\mathscr{F}$, the sequences $^{\ell}\mathbf{T}_i$ are also bounded in $L^2(B, \mathbb{R}^{n \times n})$. Thus, the $^{\ell}\mathbf{R}_{u^i}$ form bounded sequences in $L^2(B, \mathbb{R}^{n \times n})$ in accordance with the last inequality. By Hilbert's selection theorem and Rellich's embedding theorem we find a subsequence, again denoted by $^{\ell}\mathbf{R}$, which converges as follows:

$$^{\ell}\mathbf{R}_{u^i} \rightharpoonup \mathbf{R}_{u^i} \quad \text{weakly in } L^2(B, \mathbb{R}^{n \times n}), \quad {}^{\ell}\mathbf{R} \to \mathbf{R} \quad \text{strongly in } L^2(B, SO(n))$$

with some $\mathbf{R} \in H^{1,2}(B, SO(n))$. In particular, going if necessary to a subsequence, we have $^{\ell}\mathbf{R} \to \mathbf{R}$ a.e. on B as well as

$$\lim_{\ell \to \infty} \iint_B |^{\ell}\mathbf{R} \circ \widetilde{\mathbf{T}}_i - \mathbf{R} \circ \widetilde{\mathbf{T}}_i|^2 \, du \, dv = 0$$

according to the dominated convergence theorem. Hence we compute in the limit

---

[7]Note that the proof of this identity remains true for $\mathbf{R} \in H^{2,1}(B, SO(n)) \cap H^{1,2}(B, SO(n))$.

$$\lim_{\ell \to \infty} \iint_B \langle \mathbf{R} \circ \widetilde{\mathbf{T}}_i, {}^\ell\mathbf{R}_{u^i} \rangle \, du \, dv$$

$$= \lim_{\ell \to \infty} \left(\iint_B \langle \mathbf{R} \circ \widetilde{\mathbf{T}}_i - \mathbf{R} \circ \widetilde{\mathbf{T}}_i, {}^\ell\mathbf{R}_{u^i} \rangle \, du \, dv + \iint_B \langle \mathbf{R} \circ \widetilde{\mathbf{T}}_i, {}^\ell\mathbf{R}_{u^i} \rangle \, du \, dv \right)$$

$$= \iint_B \langle \mathbf{R} \circ \widetilde{\mathbf{T}}_i, \mathbf{R}_{u^i} \rangle \, du \, dv.$$

In addition, we obtain

$$\lim_{\ell \to \infty} \iint_B |{}^\ell\mathbf{R}_{u^i}|^2 \, du \, dv \geq \iint_B |\mathbf{R}_{u^i}|^2 \, du \, dv$$

due to the semicontinuity of the L^2-norm w.r.t. weak convergence. Putting the last two relations into the identity

$$|{}^\ell\mathbf{T}_i|^2 = |{}^\ell\mathbf{R}_{u^i}|^2 + 2\langle \mathbf{R} \circ \widetilde{\mathbf{T}}_i, {}^\ell\mathbf{R}_{u^i} \rangle + |\widetilde{\mathbf{T}}_i|^2$$

we finally infer

$$\lim_{\ell \to \infty} \mathscr{F}({}^\ell\mathbf{R}) = \lim_{\ell \to \infty} \iint_B \left(|{}^\ell\mathbf{T}_1|^2 + |{}^\ell\mathbf{T}_2|^2 \right) du \, dv$$

$$\geq \iint_B \left(|\mathbf{R}_u|^2 + |\mathbf{R}_v|^2 \right) du \, dv + 2 \iint_B \left(\langle \mathbf{R} \circ \widetilde{\mathbf{T}}_1, \mathbf{R}_u \rangle \right.$$

$$\left. + \langle \mathbf{R} \circ \widetilde{\mathbf{T}}_2, \mathbf{R}_v \rangle \right) du \, dv + \iint_B \left(|\widetilde{\mathbf{T}}_1|^2 + |\widetilde{\mathbf{T}}_2|^2 \right) du \, dv$$

$$= \iint_B \left(|\mathbf{T}_1|^2 + |\mathbf{T}_2|^2 \right) du \, dv$$

$$= \mathscr{F}(\mathbf{R})$$

where $\mathbf{T}_i = (T_{\sigma,i}^\vartheta)_{\sigma,\vartheta=1,\ldots,n}$ denote the torsion coefficients of the ONF $\mathfrak{N}$ with entries

$$N_\sigma := \sum_{\vartheta=1}^n R_\sigma^\vartheta \widetilde{N}_\vartheta \,.$$

Consequently, $\mathfrak{N} \in H^{1,2}(B) \cap L^\infty(B)$ minimizes $\mathscr{T}[\mathfrak{N}]$ and, in particular, it represents a weak normal Coulomb frame. $\qquad\square$

4.8.9 $H_{loc}^{2,1}$-Regularity of Weak Normal Coulomb Frames

To prove higher regularity of normal Coulomb frames we make essential use of techniques from harmonic analysis. Consult eventually our foregoing brief overview from Sects. 4.8.1 to 4.8.7 and the references we gave there. Recall that we always use conformal parameters $(u, v) \in \overline{B}$.

Proposition 4.11. *A weak normal Coulomb frame* $\mathfrak{N} \in H^{1,2}(B) \cap L^\infty(B)$ *belongs to the regularity class* $H_{loc}^{2,1}(B)$.

Proof. 1. The torsion coefficients $T_{\sigma,i}^\vartheta$ of the normal Coulomb frame $\mathfrak{N}$ are weak solutions of the Euler–Lagrange equations

$$\text{div}\,(T_{\sigma,1}^\vartheta, T_{\sigma,2}^\vartheta) = 0 \quad \text{in } B$$

for all $\sigma, \vartheta = 1, \ldots, n$. Hence, by a weak version of Poincare's lemma (see e.g. [13], Lemma 3), there exist integral functions $\tau^{(\sigma\vartheta)} \in H^{1,2}(B, \mathbb{R})$ satisfying

$$\partial_u \tau^{(\sigma\vartheta)} = -T_{\sigma,2}^\vartheta, \quad \partial_v \tau^{(\sigma\vartheta)} = T_{\sigma,1}^\vartheta \quad \text{in } B$$

in weak sense.

Thus, the weak form of the Euler–Lagrange equations can be rewritten as

$$0 = \iint_B \{\varphi_u \partial_v \tau^{(\sigma\vartheta)} - \varphi_v \partial_u \tau^{(\sigma\vartheta)}\}\, du\, dv = \int_{\partial B} \tau^{(\sigma\vartheta)} \frac{\partial\varphi}{\partial t}\, ds, \quad \varphi \in C^\infty(\overline{B}, \mathbb{R}),$$

where $\frac{\partial\varphi}{\partial t}$ denotes the tangential derivative of φ along ∂B. Note that $\tau^{(\sigma\vartheta)}|_{\partial B}$ means the L^2-trace of $\tau^{(\sigma\vartheta)}$ on the boundary curve ∂B (see e.g. the textbook Alt [2], Chap. 6, Appendix A6.6).[8] Consequently, the lemma of DuBois–Reymond[9] yields $\tau^{(\sigma\vartheta)} \equiv \text{const}$ on ∂B, and by an ordinary translation we arrive at the boundary conditions

$$\tau^{(\sigma\vartheta)} = 0 \quad \text{on } \partial B.$$

2. Thus, the integral functions $\tau^{(\sigma\vartheta)}$ are weak solutions of the second-order system

$$\Delta\tau^{(\sigma\vartheta)} = -\partial_u T_{\sigma,2}^\vartheta + \partial_v T_{\sigma,1}^\vartheta = -N_{\sigma,v} \cdot N_{\vartheta,u} + N_{\sigma,u} \cdot N_{\vartheta,v} \quad \text{in } B$$

---

[8]Let $1 \leq p \leq \infty$. Then there is a uniquely determined map $S\colon H^{1,p}(B) \to L^p(\partial B)$ such that $\|S(\phi)\|_{L^p(\partial B)} \leq C\|\phi\|_{H^{1,2}(B)}$. Additionally, it holds $S(\phi) = \phi|_{\partial B}$ if $\phi \in H^{1,2}(B, \mathbb{R}) \cap C^0(\overline{B}, \mathbb{R})$. The map S is called the *trace mapping*.

[9]Its one-dimensional version is the following: Let $f \in L^1([a, b], \mathbb{R})$ and $\int_a^b f(x)\varphi'(x)\, dx = 0$ for all $\varphi \in C^\infty([a, b], \mathbb{R})$. Then $f \equiv \text{const}$ almost everywhere.

where the second identity follows by direct differentiation. By a result of Coifman et al. [26],[10] the right-hand side of this div-curl type equation belongs to the Hardy space $\mathcal{H}^1_{loc}(B)$ and, hence, the $\tau^{(\sigma\vartheta)}$ belong to $H^{2,1}_{loc}(B,\mathbb{R})$ by Fefferman and Stein [45].[11] Consequently we find

$$T^{\vartheta}_{\sigma,i} \in H^{1,1}_{loc}(B,\mathbb{R}) \cap L^2(B,\mathbb{R}).$$

We next employ the Weingarten equations

$$N_{\sigma,u^i} = -\sum_{j,k=1}^{2} L_{\sigma,ij}\, g^{jk}\, X_{u^k} + \sum_{\vartheta=1}^{n} T^{\vartheta}_{\sigma,i}\, N_{\vartheta}$$

in a weak form. For the coefficients of the second fundamental form we have $L_{\sigma,ij} = N_{\sigma} \cdot X_{u^i u^j}$ which leads us to $L_{\sigma,ij} \in H^{1,2}(B,\mathbb{R})$ taking $X_{u^i u^j} \in L^{\infty}(B,\mathbb{R}^{n+2})$ and $\mathfrak{N} \in H^{1,2}(B) \cap L^{\infty}(B)$ into account. Hence we arrive at

$$N_{\sigma,u^i} \in H^{1,1}_{loc}(B,\mathbb{R}^{n+2}) \quad \text{and} \quad \mathfrak{N} \in H^{2,1}_{loc}(B)$$

for our weak normal Coulomb frame.

Notice that $T^{\vartheta}_{\sigma,i} \in H^{1,1}_{loc}(B,\mathbb{R}) \cap L^2(B,\mathbb{R})$ and $N_{\vartheta} \in H^{1,2}(B,\mathbb{R}^{n+2}) \cap L^{\infty}(B,\mathbb{R}^{n+2})$ imply $\sum_{\vartheta=1}^{n} T^{\vartheta}_{\sigma,i} N_{\vartheta} \in H^{1,1}_{loc}(B,\mathbb{R}^{n+2})$ by a careful adaption of the classical product rule in Sobolev spaces which is explained in the lemma below. Modulo this property the statement is proved. □

So let us come to the following lemma to complete the proof.

Lemma 4.2. *It holds*

$$\sum_{\vartheta=1}^{n} T^{\vartheta}_{\sigma,i} N_{\vartheta} \in H^{1,1}(B,\mathbb{R}^{n+2}).$$

Proof. It holds $T^{\vartheta}_{\sigma,i} N_{\vartheta} \in L^2(B,\mathbb{R}^{n+2})$ since $T^{\vartheta}_{\sigma,i} \in L^2(B,\mathbb{R})$ and $N_{\vartheta} \in L^{\infty}(B,\mathbb{R}^{n+2})$. We show that $T^{\vartheta}_{\sigma,i} N_{\vartheta}$ has a weak derivative, i.e. we prove that

$$T^{\vartheta}_{\sigma,i} \partial_{u^j} N_{\vartheta} + N_{\vartheta} \partial_{u^j} T^{\vartheta}_{\sigma,i} \in L^1(B,\mathbb{R}^{n+2})$$

is the weak derivative of $T^{\vartheta}_{\sigma,i} N_{\vartheta}$. In other words

---

[10]Let $\phi \in H^{1,2}(\mathbb{R}^2,\mathbb{R})$. Then $f := \det(\nabla\phi) \in \mathcal{H}^1(\mathbb{R}^2)$, where $\|f\|_{\mathcal{H}^1(\mathbb{R}^2)} \leq C\|\phi\|_{H^{1,2}(\mathbb{R}^2)}$; see Sect. 4.8.5 above.

[11]Let $\phi \in L^1(\mathbb{R}^2,\mathbb{R})$ be a solution of $-\Delta\phi = f \in \mathcal{H}^1(\mathbb{R}^2)$. Then all second derivatives of ϕ belong to $L^1(\mathbb{R}^2,\mathbb{R})$, and it holds $\|\phi_{x^\alpha x^\beta}\|_{L^1(\mathbb{R}^m)} \leq C\|f\|_{L^1(\mathbb{R}^m)}$ for all $\alpha,\beta = 1,2$.

$$\iint\limits_B (T_{\sigma,i}^\vartheta \partial_{u^j} N_\vartheta + N_\vartheta \partial_{u^j} T_{\sigma,i}^\vartheta)\varphi \, dudv = -\iint\limits_B (T_{\sigma,i}^\vartheta N_\vartheta)\varphi_{u^j} \, dudv$$

for all $\varphi \in C_0^\infty(B,\mathbb{R})$. For such a test function φ define

$$\psi = T_{\sigma,i}^\vartheta \varphi \in H_0^{1,1}(B,\mathbb{R}) \cap L^2(B,\mathbb{R}).$$

We want to verify

$$\iint\limits_B N_{\vartheta,u^j} \psi \, dudv = -\iint\limits_B N_\vartheta \psi_{u^j} \, dudv.$$

For the proof we approximate ψ with smooth functions $\psi^\varepsilon \in C_0^\infty(B,\mathbb{R})$ in the sense of Friedrichs. Then $\psi^\varepsilon \to \psi$ in $H^{1,1}(B,\mathbb{R}) \cap L^2(B,\mathbb{R})$, and $\psi = 0$ outside some compact set $K \subset\subset B$. We now estimate as follows

$$\left| \iint\limits_B (N_{\vartheta,u^j} \psi + N_\vartheta \psi_{u^j}) \, dudv \right|$$

$$= \left| \iint\limits_B (N_{\vartheta,u^j} \psi^\varepsilon + N_\vartheta \psi_{u^j}^\varepsilon) \, dudv \right| + \left| \iint\limits_B N_{\vartheta,u^j} (\psi - \psi^\varepsilon) \, dudv \right|$$

$$+ \left| \iint\limits_B N_\vartheta (\psi_{u^j}^\varepsilon - \psi_{u^j}) \, dudv \right|$$

$$\leq \|N_{\vartheta,u^j}\|_{L^2(B)} \|\psi - \psi^\varepsilon\|_{L^2(B)} + \|N_\vartheta\|_{L^\infty(B)} \|\psi_{u^j}^\varepsilon - \psi_{u^j}\|_{L^1(B)}$$

taking

$$\iint\limits_B N_{\vartheta,u^j} \psi^\varepsilon \, dudv = -\iint\limits_B N_\vartheta \psi_{u^j}^\varepsilon \, dudv$$

into account. Because $\|\psi - \psi^\varepsilon\|_{L^2(B)} \to 0$ and $\|\psi_{u^j}^\varepsilon - \psi_{u^j}\|_{L^1(B)} \to 0$ for $\varepsilon \to 0$ we arrive at the identity stated above. Now we use the product rule and calculate

$$\iint\limits_B (T_{\sigma,i}^\vartheta N_{\vartheta,u^j} + N_\vartheta \partial_{u^j} T_{\sigma,i}^\vartheta)\varphi \, dudv$$

$$= \iint\limits_B N_{\vartheta,u^j} \psi \, dudv + \iint\limits_B N_\vartheta \partial_{u^j} T_{\sigma,i}^\vartheta \varphi \, dudv$$

$$= - \iint_B (T^\vartheta_{\sigma,i} \varphi)_{u^j} N_\vartheta \, du \, dv + \iint_B N_\vartheta \partial_{u^j} T^\vartheta_{\sigma,i} \varphi \, du \, dv$$

$$= - \iint_B \partial_{u^j} T^\vartheta_{\sigma,i} N_\vartheta \varphi \, du \, dv - \iint_B T^\vartheta_{\sigma,i} \varphi_{u^j} N \, du \, dv + \iint_B N_\vartheta \partial_{u^j} T^\vartheta_{\sigma,i} \varphi \, du \, dv$$

$$= - \iint_B (T^\vartheta_{\sigma,i} N_\vartheta) \varphi_{u^j} \, du \, dv.$$

This proves the lemma. $\square$

4.9 Classical Regularity of Normal Coulomb Frames

Our main result of this chapter is the proof of classical regularity of normal Coulomb frames. An essential tool on our road to regularity are again the Weingarten equations from the first chapter.

We present the next proof without giving detailed references.

Theorem 4.3. *For a conformally parametrized immersion* $X \in C^{k,\alpha}(\overline{B}, \mathbb{R}^{n+2})$ *with* $k \geq 4$ *and* $\alpha \in (0, 1)$ *there exists a normal Coulomb frame*

$$\mathfrak{N} \in C^{k-1,\alpha}(\overline{B})$$

minimizing $\mathscr{T}[\mathfrak{N}]$.

Proof. 1. We fix some ONF $\widetilde{\mathfrak{N}} \in C^{k-1,\alpha}(\overline{B})$ and construct a weak normal Coulomb frame $\mathfrak{N} \in W^{1,2}(B) \cap L^\infty(B)$. We then know $\mathfrak{N} \in H^{2,1}_{loc}(B)$. Defining the orthogonal mapping $\mathbf{R} = (R^\vartheta_\sigma)_{\sigma,\vartheta=1,\dots,n}$ by $R^\vartheta_\sigma := N_\sigma \cdot \widetilde{N}_\vartheta$ we thus find

$$N_\sigma = \sum_{\vartheta=1}^n R^\vartheta_\sigma \widetilde{N}_\vartheta \quad \text{and} \quad \mathbf{R} \in H^{2,1}_{loc}(B, SO(n)) \cap H^{1,2}(B, SO(n)).$$

In particular, we can assign a curvature tensor (in matrix form)

$$\mathbf{S}_{12} = (S^\vartheta_{\sigma,12})_{\sigma,\vartheta=1,\dots,n} \in L^1_{loc}(B, \mathbb{R}^{n \times n})$$

to $\mathfrak{N}$ by formula

$$S^\vartheta_{\sigma,12} = \partial_v T^2_{\sigma,1} - \partial_u T^\vartheta_{\sigma,2} + \sum_{\omega=1}^n (T^\omega_{\sigma,1} T^\vartheta_{\omega,2} - T^\omega_{\sigma,2} T^\vartheta_{\omega,1}).$$

Moreover, from Sect. 1.6.5 we infer $\mathbf{S}_{12} \in L^\infty(B, \mathbb{R}^{n \times n})$.

2. Introduce the matrix $\tau = \left(\tau^{(\sigma\vartheta)}\right)_{\sigma,\vartheta=1,\ldots,n} \in H^{1,2}(B, \mathbb{R}^{n\times n})$ by

$$\partial_u \tau^{(\sigma\vartheta)} = -T^{\vartheta}_{\sigma,2}, \quad \partial_v \tau^{(\sigma\vartheta)} = T^{\vartheta}_{\sigma,1} \quad \text{in } B,$$
$$\tau^{(\sigma\vartheta)} = 0 \quad \text{on } \partial B.$$

Then the definition of the normal curvature tensor gives us the non-linear system

$$\Delta \tau^{\vartheta}_{\sigma} = -\sum_{\omega=1}^{n} \left(\partial_u \tau^{(\sigma\omega)} \partial_v \tau^{(\omega\vartheta)} - \partial_v \tau^{(\sigma\omega)} \partial_u \tau^{(\omega\vartheta)}\right) + S^{\vartheta}_{\sigma,12}$$

in B together with $\tau^{(\sigma\vartheta)} = 0$ on ∂B. On account of $\mathbf{S}_{12} = (S^{\vartheta}_{\sigma,12})_{\sigma,\vartheta=1,\ldots,n} \in L^{\infty}(B, \mathbb{R}^{n\times n})$, a part of Wente's inequality yields $\tau \in C^0(\overline{B}, \mathbb{R}^{n\times n})$, see e.g. Brezis and Coron [15]; compare also Rivière [98] and the corresponding boundary regularity theorem in Müller and Schikorra [91] for more general results. By appropriate reflection of τ and $\mathbf{S}_{12}$ (the reflected quantities are again denoted by τ and $\mathbf{S}_{12}$) we obtain a weak solution $\tau \in W^{1,2}(B_{1+d}, \mathbb{R}^{n\times n}) \cap C^0(B_{1+d}, \mathbb{R}^{n\times n})$ of

$$\Delta \tau = f(w, \nabla\tau) \quad \text{in } B_{1+d} := \{w \in \mathbb{R}^2 : |w| < 1 + d\}$$

with some $d > 0$ and a right-hand side f satisfying

$$|f(w, p)| \le a|p|^2 + b \quad \text{for all } p \in \mathbb{R}^{2n^2}, \quad w \in B_{1+d},$$

with some reals $a, b > 0$. Now, applying Tomi's regularity result from [115] for weak solutions of this non-linear system possessing small variation locally in B_{1+d}, we find $\tau \in C^{1,\nu}(\overline{B}, \mathbb{R}^{n\times n})$ for any $\nu \in (0,1)$ (notice that Tomi's result applies for such systems with $b = 0$, but his proof can easily be adapted to our inhomogeneous case $b > 0$).

3. From the first-order system for $\tau^{(\sigma\vartheta)}$ we infer $\mathbf{T}_i \in C^{\alpha}(\overline{B}, \mathbb{R}^{n\times n})$. Thus, the Weingarten equations for N_{σ,u^i} yield

$$\mathfrak{N} \in W^{1,\infty}(B)$$

on account of $\mathfrak{N} \in L^{\infty}(B)$, and we obtain from Sobolev's embedding theorem

$$\mathfrak{N} \in C^{\alpha}(\overline{B}).$$

Inserting this again into the Weingarten equations, we find

$$\mathfrak{N} \in C^{1,\alpha}(\overline{B}).$$

Hence, we can conclude $\mathbf{R} = (N_\sigma \cdot \widetilde{N}_\vartheta)_{\sigma,\vartheta=1,\ldots,n} \in C^{1,\alpha}(\overline{B}, \mathbb{R}^{n \times n})$, and the transformation rule $\mathbf{S}_{12} = \mathbf{R} \circ \widetilde{\mathbf{S}}_{12} \circ \mathbf{R}^t$ from Sect. 1.6.5 implies

$$\mathbf{S}_{12} = \left(S^\vartheta_{\sigma,12}\right)_{\sigma,\vartheta=1,\ldots,n} \in C^\alpha(\overline{B}, \mathbb{R}^{n \times n})$$

Now the right-hand side of the equation for $\Delta \tau^{(\sigma\vartheta)}$ belongs to $C^\alpha(\overline{B}, \mathbb{R})$, and potential theoretic estimates ensure $\tau \in C^{2,\alpha}(\overline{B}, \mathbb{R}^{n \times n})$. Involving again our first-order system for the $\tau^{(\sigma\vartheta)}$ gives $\mathbf{T}_i \in C^{1,\alpha}(\overline{B}, \mathbb{R}^{n \times n})$, which proves

$$\mathfrak{N} \in C^{2,\alpha}(\overline{B})$$

using the Weingarten equations once more. Finally, for $k \geq 4$, we can bootstrap by concluding $\mathbf{R} \in C^{2,\alpha}(\overline{B}, \mathbb{R}^{n \times n})$ and $\mathbf{S}_{12} \in C^{1,\alpha}(\overline{B}, \mathbb{R}^{n \times n})$ from the transformation rule for $\mathbf{S}_{12}$ and repeating the arguments above.
The proof is complete.   □

References

1. G. Albrecht, The Veronese surface revisited. J. Geom. **73**, 22–38 (2002)
2. W. Alt, *Lineare Funktionalanalysis* (Springer, New York, 2006)
3. B. Andrews, C. Baker, Mean curvature flow of pinched submanifolds to spheres. J. Differ. Geom. **85**(3), 357–396 (2010)
4. C. Bär, *Elementare Differentialgeometrie* (Walter de Gruyter GmbH & Co. KG, Berlin, 2010)
5. J.L. Barbosa, M. do Carmo, On the size of stable minimal surfaces in $\mathbb{R}^3$. Am. J. Math. **98**, 515–528 (1974)
6. J.L. Barbosa, M. do Carmo, Stability of minimal surfaces and eigenvalues of the Laplacian. Math. Z. **173**, 13–28 (1980)
7. H.W. Begehr, *Complex Analytic Methods for Partial Differential Equations* (World Scientific Publishing, River Edge, 1994)
8. M. Bergner, S. Fröhlich, On two-dimensional immersions of prescribed mean curvature in $\mathbb{R}^n$. Z. Anal. Anw. **27**(1), 31–52 (2008)
9. M. Bergner, R. Jakob, Exclusion of boundary branch points for minimal surfaces. Analysis **31**, 181–190 (2011)
10. S. Bernstein, Über ein geometrisches Theorem und seine Anwendung auf die partiellen Differentialgleichungen vom elliptischen Typus. Math. Z. **26**, 551–558 (1927)
11. L. Bers, Univalent solutions of linear elliptic systems. Comm. Pure Appl. Math. **VI**, 513–526 (1953)
12. W. Blaschke, K.Leichtweiss, *Elementare Differentialgeometrie* (Springer, New York, 1973)
13. J. Bourgain, H. Brezis, P. Mironescu, Lifting in Sobolev spaces. J. Anal. Math. **80**, 37–86 (2000)
14. H. Brauner, *Differentialgeometrie* (Friedrich Vieweg & Sohn Verlagsgesellschaft mbH, Braunschweig, 1981)
15. H. Brezis, J.M. Coron, Multiple solutions of H-systems and Rellich's conjecture. Comm. Pure Appl. Math. **37**, 149–187 (1984)
16. R. Bryant, A duality theorem for Willmore surfaces. J. Differ. Geom. **20**, 23–53 (1984)
17. J.L. Buchanan, A similarity principle for Pascali systems. Complex Variables **1**, 155–165 (1983)
18. J.L. Buchanan, R.P. Gilbert, *First Order Elliptic Systems* (Academic, New York, 1983)
19. A. Burchard, L.E. Thomas, On the Cauchy problem for a dynamical Euler's elastica. Comm. Part. Differ. Equat. **28**, 271–300 (2003)
20. H. Cartan, *Differential forms* (Dover publications, Mineola, New York, 2006)
21. B.Y. Chen, *Geometry of Submanifolds* (Marcel Dekker, New York, 1973)
22. B.Y. Chen, in *Riemannian Submanifolds*. Handbook of Differential Geometrie 1 (Elsevier Science B.V., The Netherlands, 1999), pp. 187–418

S. Fröhlich, *Coulomb Frames in the Normal Bundle of Surfaces in Euclidean Spaces*,   107
Lecture Notes in Mathematics 2053, DOI 10.1007/978-3-642-29846-2,
© Springer-Verlag Berlin Heidelberg 2012

23. B.Y. Chen, G.D. Ludden, Surfaces with mean curvature vector parallel in the normal bundle. Nagoya Math. J. **47**, 161–167 (1972)
24. S.S. Chern, An elementary proof of the existence of isothermal parameters on a surface. Proc. Am. Math. Soc. **6**(5), 771–782 (1955)
25. M.A. Cheshkova, Evolute surfaces in E^4. Math. Notes **70**(6), 870–872 (2001)
26. R. Coifman, P.L. Lions, Y. Meyer, S. Semmes, Compensated compactness and Hardy spaces. J. Math. Pures Appl. **9**(3), 247–286 (1993)
27. T.H. Colding, W.P. Minicozzi, *Minimal Surfaces* (Courant Institute of Mathematical Sciences, New York, 1999)
28. R. Courant, *Dirichlet's Principle, Conformal Mappings, and Minimal Surfaces* (Interscience publishing, New York, 1950)
29. R. Courant, D. Hilbert, *Methods of Mathematical Physics 2* (Wiley, New York, 1962)
30. R.C.T. da Costa, Constraints in quantum mechanics. Phys. Rev. **A25**(6), 2893–2900 (1982)
31. A. Dall'Acqua, Uniqueness for the homogeneous Dirichlet Willmore boundary value problem (2012). Ann. Glob. Anal. Geom. DOI: 10.1007/S10455-012-9320-6
32. A. Dall'Acqua, K. Deckelnick, H.-Chr. Grunau, Classical solutions to the Dirichlet problem for Willmore surfaces of revolution. Adv. Calc. Var. **1**, 379–397 (2008)
33. U. Dierkes, Maximum principles for submanifolds of arbitrary codimension and bounded mean curvature. Calc. Var. **22**, 173–184 (2005)
34. U. Dierkes, S. Hildebrandt, F. Sauvigny, *Minimal Surfaces* (Springer, New York, 2010)
35. P.M. Dirac, *Lectures on Quantum Mechanics* (Dover Publications, New York, 2001)
36. M.P. do Carmo, *Riemannian Geometry* (Birkhäuser, Boston, 1992)
37. M. Dobrowolski, *Angewandte Funktionalanalysis* (Springer, New York, 2006)
38. H. Dorn, G. Jorjadze, S. Wuttke, On spacelike and timelike minimal surfaces in AdSn. JHEP 05 (2009) 048. DOI: 10.1088/1126-6708/2009/05/048
39. P. Duren, *Theory of H^p Spaces* (Dover Publications, New York, 2000)
40. P. Duren, *Harmonic Mappings in the Plane* (Cambridge University Press, Cambridge, 2004)
41. K. Ecker, *Regularity Theory for Mean Curvature Flow* (Birkhäuser, Boston, 2004)
42. K. Ecker, G. Huisken, Interior curvature estimates for hypersurfaces of prescribed mean curvature. Ann. Inst. H. Poincaré Anal. Non Linéaire **6**, 251–260 (1989)
43. L.P. Eisenhart, *Riemannian Geometry* (Princeton University Press, Princeton, 1949)
44. J.-H. Eschenburg, J. Jost, *Differentialgeometrie und Minimalflächen* (Springer, New York, 2007)
45. C. Fefferman, E.M. Stein, H^p spaces of several variables. Acta Math. **129**, 137–193 (1972)
46. S. Fröhlich, Curvature estimates for μ-stable G-minimal surfaces and theorems of Bernstein type. Analysis **22**, 109–130 (2002)
47. S. Fröhlich, Katenoidähnliche Lösungen geometrischer Variationsprobleme (2004). Preprint 2322, FB Mathematik, TU Darmstadt
48. S. Fröhlich, On 2-surfaces in $\mathbb{R}^4$ and $\mathbb{R}^n$. in *Proceedings of the 5th Conference of Balkan Society of Geometers*, Mangalia, 2005
49. S. Fröhlich, F. Müller, On critical normal sections for two-dimensional immersions in $\mathbb{R}^4$ and a Riemann–Hilbert problem. Differ. Geom. Appl. **26**, 508–513 (2008)
50. S. Fröhlich, F. Müller, On critical normal sections for two-dimensional immersions in $\mathbb{R}^{n+2}$. Calc. Var. **35**, 497–515 (2009)
51. S. Fröhlich, F. Müller, On the existence of normal Coulomb frames for two-dimensional immersions with higher codimension. Analysis **31**, 221–236 (2011)
52. S. Fröhlich, S. Winklmann, Curvature estimates for graphs with prescribed mean curvature and flat normal bundle. Manuscripta Math. **122**(2), 149–162 (2007)
53. D. Gilbarg, N.S. Trudinger, *Elliptic Partial Differential Equations of Second Order* (Springer, New York, 1983)
54. K. Große-Brauckmann, R. Kusner, J. Sullivan, Coplanar constant mean curvature surfaces. Comm. Anal. Geom. **5**, 985–1023 (2007)
55. I.V. Guadalupe, L. Rodriguez, Normal curvature of surfaces in space forms. Pac. J. Math. **106**(1), 95–103 (1983)

56. R. Gulliver, in *Minimal Surfaces of Finite Index in Manifolds of Positive Scalar Curvature*, ed. by S. Hildebrandt, D. Kinderlehrer, M. Miranda. Calculus of Variations and Partial Differential Equations (Springer, New York, 1988)
57. R.R. Hall, On an inequality of E. Heinz. J. Anal. Math. **42**, 185–198 (1982)
58. P. Hartman, A. Wintner, On the existence of Riemannian manifolds which cannot carry non-constant analytic or harmonic functions in the small. Am. J. Math. **75**, 260–276 (1953)
59. P. Hartman, A. Wintner, On the local behaviour of solutions of nonparabolic partial differential equations. Am. J. Math. **75**, 449–476 (1953)
60. E. Heil, *Differentialformen und Anwendungen auf Vektoranalysis, Differentialgleichungen, Geometrie* (Bibliographisches Institut, Germany, 1974)
61. E. Heinz, Über die Lösungen der Minimalflächengleichung. Nachr. Akad. Gött., Math.-Phys. Kl., 51–56 (1952)
62. E. Heinz, On certain nonlinear elliptic differential equations and univalent mappings. J. Anal. Math. **5**, 197–272 (1957)
63. E. Heinz, Über das Randverhalten quasilinearer elliptischer Systeme mit isothermen Parametern. Math. Z. **113**, 99–105 (1970)
64. F. Helein, *Harmonic Maps, Conservation Laws and Moving Frames* (Cambridge University Press, Cambridge, 2002)
65. W. Helfrich, Elastic properties of lipid bilayers: Theory and possible experiments. Z. Naturforsch. **28c**, 693–703 (1973)
66. S. Hildebrandt, Einige Bemerkungen über Flächen beschränkter mittlerer Krümmung. Math. Z. **115**, 169–178 (1970)
67. S. Hildebrandt, H. von der Mosel, On Lichtenstein's theorem about globally conformal mappings. Calc. Var. **23**, 415–424 (2005)
68. S. Hildebrandt, J. Jost, K.O. Widman, Harmonic mappings and minimal submanifolds. Invent. Math. **62**, 269–298 (1980)
69. D. Hoffman, R. Osserman, *The Geometry of the Generalized Gauss Map* (American Mathematical Society, Providence, 1980)
70. E. Hopf, Bemerkungen zu einem Satze von S. Bernstein aus der Theorie der elliptischen Differentialgleichungen. Math. Z. **29**, 744–745 (1929)
71. E. Hopf, On S. Bernstein's theorem on surfaces $z(x, y)$ of nonpositive curvature. Proc. Am. Math. Soc. **1**(1), 80–85 (1950)
72. H. Hopf, Über Flächen mit einer Relation zwischen den Hauptkrümmungen. Math. Nachr. **4**, 232–249 (1950/51)
73. R. Jakob, Finiteness of the set of solutions of Plateau's problem for polygonal boundary curves. Ann. I.H. Poincaré Anal. Non Linéaire **24**, 963–987 (2007)
74. R. Jakob, About the finiteness of the set of solutions of Plateau's problem for polygonal boundary curves. Analysis **29**, 365–385 (2009)
75. R. Jakob, Finiteness of the number of solutions of Plateau's problem for polygonal boundary curves II. Ann. Global Anal. Geom. **36**, 19–35 (2009)
76. J. Jost, Y.L. Xin, Bernstein type theorems for higher codimension. Calc. Var. **9**, 277–296 (1999)
77. H. Kalf, T. Kriecherbauer, E. Wienholtz, *Elliptische Differentialgleichungen Zweiter Ordnung* (Springer, New York, 2009)
78. K. Kenmotsu, *Surfaces of Constant Mean Curvature* (Oxford University Press, Oxford, 2003)
79. K. Kenmotsu, D. Zhou, The classification of the surfaces with parallel mean curvature vector in two-dimensional complex space forms. Am. J. Math. **122**, 295–317 (2000)
80. W. Klingenberg, *Eine Vorlesung über Differentialgeometrie* (Springer, New York, 1973)
81. K. Kobayashi, Fundamental equations for submanifolds. Fortschr. Phys. **37**, 599–610 (1989)
82. B.G. Konopelchenko, G. Landolfi, On rigid string instantons in four dimensions. Phys. Lett. **B459**, 522–526 (1999)
83. A. Korn, in *Zwei Anwendungen der Methode der sukzessiven Approximation*. Mathematische Abhandlungen Hermann Amandus Schwarz zu seinem fünfzigjährigem Doktorjubiläum am 6. August 1914 gewidmet von Freunden und Schülern (Springer, New York, 1914), pp. 215–229

84. W. Kühnel, *Differentialgeometrie* (Friedrich Vieweg & Sohn Verlagsgesellschaft mbH, Braunschweig, 1999)
85. H.B. Lawson, Complete minimal surfaces in S^3. Ann. Math. **92**, 335–374 (1970)
86. H. Li, Willmore surfaces in S^n. Ann. Glob. Anal. Geom. **21**, 203–213 (2002)
87. L. Lichtenstein, Zur Theorie der konformen Abbildung. Konforme Abbildung nichtanalytischer, singularitätenfreier Flächenstücke auf ebene Gebiete. Bull. Int. de l'Acad. Sci. Cracovie **A**, 192–217 (1916)
88. M.J. Micallef, Stable minimal surfaces in Euclidean space. J. Differ. Geom. **19**, 57–84 (1984)
89. R. Moser, *Partial Regularity, Harmonic Maps and Related Problems* (World Scientific Publishing, River Edge, 2005)
90. F. Müller, Funktionentheorie und Minimalflächen. Lecture notes, University Duisburg (2011)
91. F. Müller, A. Schikorra, Boundary regularity via Uhlenbeck-Rivière decomposition. Analysis **29**(2), 199–220 (2009)
92. J.C.C. Nitsche, *Vorlesungen über Minimalflächen* (Springer, New York, 1975)
93. R. Osserman, Global properties of minimal surfaces in $\mathbb{R}^3$ and $\mathbb{R}^n$. Ann. Math. **80**(2), 340–364 (1964)
94. R. Osserman, *A Survey of Minimal Surfaces* (Dover Publications, New York, 1986)
95. B. Palmer, Uniqueness theorems for Willmore surfaces with fixed and free boundaries. Indiana Univ. Math. J. **49**, 1581–1601 (2000)
96. M. Pinl, Abwickelbare Schiebflächen in R_n. Comm. Math. Helv. **24**, 64–67 (1950)
97. M. Pinl, B-Kugelbilder reeller Minimalflächen in R_4. Math. Z. **59**, 290–295 (1953)
98. T. Rivière, Conservation laws for conformally invariant variational problems. Invent. Math. **168**, 1–22 (2007)
99. T. Rivière, Analysis aspects of Willmore surfaces. Invent. math. **174**, 1–45 (2008)
100. H. Ruchert, Ein Eindeutigkeitssatz für Flächen konstanter mittlerer Krümmung. Arch. math. **33**, 91–104 (1979)
101. K. Sakamoto, Variational problems of normal curvature tensor and concircular scalar fields. Tohoku Math. J. **55**, 207–254 (2003)
102. F. Sauvigny, Die zweite Variation von Minimalflächen im $\mathbb{R}^p$ mit polygonalem Rand. Math. Z. **189**, 167–184 (1985)
103. F. Sauvigny, Ein Eindeutigkeitssatz für Minimalflächen im $\mathbb{R}^p$ mit polygonalem Rand. J. Reine Angew. Math. **358**, 92–96 (1985)
104. F. Sauvigny, On the Morse index of minimal surfaces in $\mathbb{R}^p$ with polygonal boundaries. Manuscripta Math. **53**, 167–197 (1985)
105. F. Sauvigny, A-priori-Abschätzungen der Hauptkrümmungen für Immersionen vom Mittleren-Krümmungs-Typ mittels Uniformisierung und Sätze vom Bernstein-Typ. Habilitationsschrift, Göttingen (1988)
106. F. Sauvigny, On immersions of constant mean curvature: Compactness results and finiteness theorems for Plateau's problem. Arch. Rat. Mech. Anal. **110**, 125–140 (1990)
107. F. Sauvigny, *Partielle Differentialgleichungen der Geometrie und Physik* (Springer, New York, 2005)
108. R. Schätzle, The Willmore boundary value problem. Calc. Var. **37**, 275–302 (2010)
109. R. Schoen, L. Simon, S.T. Yau, Curvature estimates for minimal hypersurfaces. Acta Math. **134**, 275–288 (1975)
110. J.A. Schouten, *Ricci Calculus* (Springer, New York, 1954)
111. F. Schulz, *Regularity Theory for Quasilinear Elliptic Systems and Monge-Ampère Equations in Two Dimensions* (Springer, New York, 1991)
112. B. Sharp, P. Topping, Decay estimates for Rivière's equation, with applications to regularity and compactness. Trans. Am. Math. Soc. (to appear)
113. K. Smoczyk, G. Wang, Y.L. Xin, Bernstein type theorems with flat normal bundle. Calc. Var. **26**, 57–67 (2006)
114. E.M. Stein, *Harmonic Analysis: Real-Variable Methods, Orthogonality, and Oscillatory Integrals* (Princeton University Press, Princeton, 1993)

115. F. Tomi, Ein einfacher Beweis eines Regularitätssatzes für schwache Lösungen gewisser elliptischer Systeme. Math. Z. **112**, 214–218 (1969)
116. P. Topping, The optimal constant in Wente's L^∞-estimate. Comment. Math. Helv. **72**, 316–328 (1997)
117. I.N. Vekua, *Verallgemeinerte Analytische Funktionen* (Akademie, Berlin, 1963)
118. A. Voss, Zur Theorie der Transformation quadratischer Differentialausdrücke und der Krümmung höherer Mannigfaltigkeiten. Math. Ann. **16**, 129–179 (1880)
119. M.-T. Wang, On graphic Bernstein type results in higher codimension. Trans. Am. Math. Soc. **355**, 265–271 (2003)
120. J.L. Weiner, On a problem of Chen, Willmore, et al. Indiana Univ. Math. J. **27**(1), 19–35 (1978)
121. W.L. Wendland, *Elliptic Systems in the Plane* (Pitman Publishing, London, 1979)
122. H.C. Wente, The differential equation $\Delta x = 2H x_u \wedge x_v$ with vanishing boundary values. Proc. Am. Math. Soc. **50**, 131–137 (1975)
123. H.C. Wente, Large solutions to the volume constrained Plateau problem. Arch. Rat. Mech. Anal. **75**, 59–77 (1980)
124. H. Weyl, Zur Infinitesimalgeometrie: p-dimensionale Fläche im n-dimensionalen Raum. Math. Z. **12**, 154–160 (1922)
125. T.J. Willmore, *Riemannian Geometry* (Oxford Science Publications, Clarendon Press, Oxford, 1993)
126. W. Wirtinger, Eine Determinantenidentität und ihre Anwendung auf analytische Gebilde und Hermitesche Massbestimmung. Monatsh. Math. Phys. **44**, 343–365 (1936)
127. Y.L. Xin, Curvature estimates for submanifolds with prescribed Gauss image and mean curvature. Calc. Var. **37**, 385–405 (2010)
128. B. Zwiebach, *A First Course in String Theory* (Cambridge University Press, Cambridge, 2004)

List of Names

S. Fröhlich, *Coulomb Frames in the Normal Bundle of Surfaces in Euclidean Spaces*,
Lecture Notes in Mathematics 2053, DOI 10.1007/978-3-642-29846-2,
© Springer-Verlag Berlin Heidelberg 2012

Index

S. Fröhlich, *Coulomb Frames in the Normal Bundle of Surfaces in Euclidean Spaces,*   115
Lecture Notes in Mathematics 2053, DOI 10.1007/978-3-642-29846-2,
© Springer-Verlag Berlin Heidelberg 2012

LECTURE NOTES IN MATHEMATICS   ⟨🐴⟩ Springer

Edited by J.-M. Morel, B. Teissier; P.K. Maini

Editorial Policy (for the publication of monographs)

1. Lecture Notes aim to report new developments in all areas of mathematics and their applications - quickly, informally and at a high level. Mathematical texts analysing new developments in modelling and numerical simulation are welcome.
   Monograph manuscripts should be reasonably self-contained and rounded off. Thus they may, and often will, present not only results of the author but also related work by other people. They may be based on specialised lecture courses. Furthermore, the manuscripts should provide sufficient motivation, examples and applications. This clearly distinguishes Lecture Notes from journal articles or technical reports which normally are very concise. Articles intended for a journal but too long to be accepted by most journals, usually do not have this "lecture notes" character. For similar reasons it is unusual for doctoral theses to be accepted for the Lecture Notes series, though habilitation theses may be appropriate.

2. Manuscripts should be submitted either online at www.editorialmanager.com/lnm to Springer's mathematics editorial in Heidelberg, or to one of the series editors. In general, manuscripts will be sent out to 2 external referees for evaluation. If a decision cannot yet be reached on the basis of the first 2 reports, further referees may be contacted: The author will be informed of this. A final decision to publish can be made only on the basis of the complete manuscript, however a refereeing process leading to a preliminary decision can be based on a pre-final or incomplete manuscript. The strict minimum amount of material that will be considered should include a detailed outline describing the planned contents of each chapter, a bibliography and several sample chapters.
   Authors should be aware that incomplete or insufficiently close to final manuscripts almost always result in longer refereeing times and nevertheless unclear referees' recommendations, making further refereeing of a final draft necessary.
   Authors should also be aware that parallel submission of their manuscript to another publisher while under consideration for LNM will in general lead to immediate rejection.

3. Manuscripts should in general be submitted in English. Final manuscripts should contain at least 100 pages of mathematical text and should always include

   - a table of contents;
   - an informative introduction, with adequate motivation and perhaps some historical remarks: it should be accessible to a reader not intimately familiar with the topic treated;
   - a subject index: as a rule this is genuinely helpful for the reader.

   For evaluation purposes, manuscripts may be submitted in print or electronic form (print form is still preferred by most referees), in the latter case preferably as pdf- or zipped psfiles. Lecture Notes volumes are, as a rule, printed digitally from the authors' files. To ensure best results, authors are asked to use the LaTeX2e style files available from Springer's web-server at:

   ftp://ftp.springer.de/pub/tex/latex/svmonot1/ (for monographs) and
   ftp://ftp.springer.de/pub/tex/latex/svmultt1/ (for summer schools/tutorials).

Additional technical instructions, if necessary, are available on request from lnm@springer.com.

4. Careful preparation of the manuscripts will help keep production time short besides ensuring satisfactory appearance of the finished book in print and online. After acceptance of the manuscript authors will be asked to prepare the final LaTeX source files and also the corresponding dvi-, pdf- or zipped ps-file. The LaTeX source files are essential for producing the full-text online version of the book (see http://www.springerlink.com/openurl.asp?genre=journal&issn=0075-8434 for the existing online volumes of LNM). The actual production of a Lecture Notes volume takes approximately 12 weeks.

5. Authors receive a total of 50 free copies of their volume, but no royalties. They are entitled to a discount of 33.3 % on the price of Springer books purchased for their personal use, if ordering directly from Springer.

6. Commitment to publish is made by letter of intent rather than by signing a formal contract. Springer-Verlag secures the copyright for each volume. Authors are free to reuse material contained in their LNM volumes in later publications: a brief written (or e-mail) request for formal permission is sufficient.

Addresses:
Professor J.-M. Morel, CMLA,
École Normale Supérieure de Cachan,
61 Avenue du Président Wilson, 94235 Cachan Cedex, France
E-mail: morel@cmla.ens-cachan.fr

Professor B. Teissier, Institut Mathématique de Jussieu,
UMR 7586 du CNRS, Équipe "Géométrie et Dynamique",
175 rue du Chevaleret
75013 Paris, France
E-mail: teissier@math.jussieu.fr

For the "Mathematical Biosciences Subseries" of LNM:

Professor P. K. Maini, Center for Mathematical Biology,
Mathematical Institute, 24-29 St Giles,
Oxford OX1 3LP, UK
E-mail : maini@maths.ox.ac.uk

Springer, Mathematics Editorial, Tiergartenstr. 17,
69121 Heidelberg, Germany,
Tel.: +49 (6221) 4876-8259

Fax: +49 (6221) 4876-8259
E-mail: lnm@springer.com